This series aims to report new developments in mathematical economics and operations research and teaching quickly, informally and at a high level. The type of material considered for publication includes:

1. Preliminary drafts of original papers and monographs

2. Lectures on a new field, or presenting a new angle on a classical field

3. Seminar work-outs

4. Reports of meetings

Texts which are out of print but still in demand may also be considered if they fall within these categories.

The timeliness of a manuscript is more important than its form, which may be unfinished or tentative. Thus, in some instances, proofs may be merely outlined and results presented which have been or will later be published elsewhere.

Publication of *Lecture Notes* is intended as a service to the international mathematical community, in that a commercial publisher, Springer-Verlag, can offer a wider distribution to documents which would otherwise have a restricted readership. Once published and copyrighted, they can be documented in the scientific literature.

Manuscripts

Manuscripts are reproduced by a photographic process; they must therefore be typed with extreme care. Symbols not on the typewriter should be inserted by hand in indelible black ink. Corrections to the typescript should be made by sticking the amended text over the old one, or by obliterating errors with white correcting fluid. Should the text, or any part of it, have to be retyped, the author will be reimbursed upon publication of the volume. Authors receive 75 free copies.

The typescript is reduced slightly in size during reproduction; best results will not be obtained unless the text on any one page is kept within the overall limit of 18 x 26.5 cm (7 x 10 ½ inches). The publishers will be pleased to supply on request special stationery with the typing area outlined.

Manuscripts in English, German or French should be sent to Prof. Dr. M. Beckmann, Department of Economics, Brown University, Providence, Rhode Island 02912/USA or Prof. Dr. H. P. Künzi, Institut für Operations Research und elektronische Datenverarbeitung der Universität Zürich, Sumatrastraße 30, 8006 Zürich.

Die *„Lecture Notes"* sollen rasch und informell, aber auf hohem Niveau, über neue Entwicklungen der mathematischen Ökonometrie und Unternehmensforschung berichten, wobei insbesondere auch Berichte und Darstellungen der für die praktische Anwendung interessanten Methoden erwünscht sind. Zur Veröffentlichung kommen:

1. Vorläufige Fassungen von Originalarbeiten und Monographien.

2. Spezielle Vorlesungen über ein neues Gebiet oder ein klassisches Gebiet in neuer Betrachtungsweise.

3. Seminarausarbeitungen.

4. Vorträge von Tagungen.

Ferner kommen auch ältere vergriffene spezielle Vorlesungen, Seminare und Berichte in Frage, wenn nach ihnen eine anhaltende Nachfrage besteht.

Die Beiträge dürfen im Interesse einer größeren Aktualität durchaus den Charakter des Unfertigen und Vorläufigen haben. Sie brauchen Beweise unter Umständen nur zu skizzieren und dürfen auch Ergebnisse enthalten, die in ähnlicher Form schon erschienen sind oder später erscheinen sollen.

Die Herausgabe der *„Lecture Notes"* Serie durch den Springer-Verlag stellt eine Dienstleistung an die mathematischen Institute dar, indem der Springer-Verlag für ausreichende Lagerhaltung sorgt und einen großen internationalen Kreis von Interessenten erfassen kann. Durch Anzeigen in Fachzeitschriften, Aufnahme in Kataloge und durch Anmeldung zum Copyright sowie durch die Versendung von Besprechungsexemplaren wird eine lückenlose Dokumentation in den wissenschaftlichen Bibliotheken ermöglicht.

Lecture Notes in Operations Research and Mathematical Systems

Economics, Computer Science, Information and Control

Edited by M. Beckmann, Providence and H. P. Künzi, Zürich

27

I. H. Mufti

National Research Council
Ottawa, Ontario/Canada

Computational Methods in Optimal Control Problems

Springer-Verlag
Berlin · Heidelberg · New York 1970

Advisory Board

H. Albach · A. V. Balakrishnan · F. Ferschl
W. Krelle · N. Wirth

ISBN-13: 978-3-540-04951-7   e-ISBN-13: 978-3-642-85960-1
DOI: 10.1007/ 978-3-642-85960-1

This work is subject to copyright. All rights are reserved, whether the whole or part of the material is concerned, specifically those of translation, reprinting, re-use of illustrations, broadcasting, reproduction by photocopying machine or similar means, and storage in data banks.

Under § 54 of the German Copyright Law where copies are made for other than private use, a fee is payable to the publisher, the amount of the fee to be determined by agreement with the publisher.

© by Springer-Verlag Berlin · Heidelberg 1970. Library of Congress Catalog Card Number 77-121990.
Title No. 3776

PREFACE

The purpose of this modest report is to present in a simplified manner some of the computational methods that have been developed in the last ten years for the solution of optimal control problems. Only those methods that are based on the minimum (maximum) principle of Pontriagin are discussed here.

The outline of the report is as follows: In the first two sections a control problem of Bolza is formulated and the necessary conditions in the form of the minimum principle are given. The method of steepest descent and a conjugate gradient-method are discussed in Section 3. In the remaining sections, the successive sweep method, the Newton-Raphson method and the generalized Newton-Raphson method (also called quasilinearization method) are presented from a unified approach which is based on the application of Newton-Raphson approximation to the necessary conditions of optimality. The second-variation method and other shooting methods based on minimizing an error function are also considered.

TABLE OF CONTENTS

1.0	INTRODUCTION	1
2.0	NECESSARY CONDITIONS FOR OPTIMALITY	2
3.0	THE GRADIENT METHOD	4
	3.1 Min H Method and Conjugate Gradient Method	8
	3.2 Boundary Constraints	9
	3.3 Problems with Control Constraints	15
4.0	SUCCESSIVE SWEEP METHOD	18
	4.1 Final Time Given Implicitly	22
5.0	SECOND-VARIATION METHOD	23
6.0	SHOOTING METHODS	27
	6.1 Newton-Raphson Method	27
	6.2 Minimizing Methods	34
7.0	THE GENERALIZED NEWTON-RAPHSON METHOD	35
8.0	CONCLUDING REMARKS	41
	REFERENCES	43

1.0 INTRODUCTION

In recent years a considerable amount of work has been done in developing numerical techniques for the solution of optimal control problems. Consider a system of differential equations and initial conditions

$$\dot{x} = f(x,t,u), \quad x(t_0) = x_0; \; t_0, \; x_0 \sim \text{given} \quad \ldots\ldots(1)$$

where x, the state vector, and f are n-vectors and u is an m-vector. The m dimensional vector u(t) which is called the control function, or control for the system may have to satisfy the constraints of the form

$$\alpha_j \leq u_j \leq \beta_j, \quad \beta_j > \alpha_j \quad j = 1,\ldots,m. \quad \ldots\ldots(2)$$

The problem of optimal control is to choose the control u(t) so as to transfer the initial state x_0 in accordance with $\dot{x} = f(x,t,u)$ to the target set given by

$$\psi(x(T),T) = 0 \quad \ldots\ldots(3)$$

where ψ is an h-vector[*], in such a way as to minimize a functional

$$J(u) = M(x(T),T) + \int_{t_0}^{T} L(x,t,u)\,dt \quad \ldots\ldots(4)$$

where M is a function defined on the set of terminal states and the terminal time and the integral is evaluated along the solution of (1). This problem, in the literature, is called a control problem of Bolza.

---

[*] $h \leq n+1$ if T is variable and $h \leq n$ if T is fixed.

Most of the numerical methods[*] that have been developed in recent times for the solution of the control problems are based either on the minimum (maximum) principle of Pontriagin (see e.g. [2] - [4]) or on the Bellman's dynamic programming (see e.g. [5] - [7]) In this paper, however, we shall discuss only those methods that are based on the minimum (maximum) principle of Pontriagin.

This paper, apart from the introductory section, is divided into seven sections. In Section 2 the Pontriagin principle is stated for the problem under consideration. Sections 3 - 7 discuss the gradient method, the successive sweep method, the second-variation method, the shooting methods and the generalized Newton-Raphson method respectively. In Section 8 some concluding remarks are made.

2.0 NECESSARY CONDITIONS FOR OPTIMALITY

Let us consider first the optimal control problem when there are no constraints of the form (3), i.e. free end point problem. Then, under appropriate conditions of continuity and differentiability on the functions involved, the following theorem can be proved ([8] and [9]).

Theorem 1 (Minimum Principle). Let $u^*(t)$ be an optimal control and let $x^*(t)$ be the corresponding optimal trajectory. Then there exist multipliers $p_0 \geq 0$, $p_1(t),\ldots,p_n(t)$, not vanishing simultaneously, such that the following conditions are fulfilled:

(a) The vector $p(t) = (p_1(t),\ldots,p_n(t))'$ is continuous on $[t_0,T^*]$ (T^* is the transfer time) and the functions $x^*(t)$ and $p(t)$ satisfy the canonical (or Hamiltonian) system of differential equations

---

[*]The classical method of Ritz has also been used in control problems (see e.g. Ref. [1]).

$$\dot{x} = \frac{\partial H}{\partial p}, \quad \dot{p} = -\frac{\partial H}{\partial x} \quad \ldots(5)$$

where H, the Hamiltonian, is given by

$$H(x,p_0,p,u,t) = p_0 L + p'f \quad (' \equiv \text{transpose}) \quad \ldots(6)$$

(b) $u^*(t)$ minimizes $H(x^*,p_0,p,u,t)$ over the admissible set U of u for $t \in [t_0, T^*]$, i.e.

$$H(x^*,p_0,p,u^*,t) \leq H(x^*,p_0,p,u,t) \quad \ldots(7)$$

for all $u \in U$.

(c) At the terminal point $(T^*, x^*(T^*))$ the transversality conditions hold:

$$p(T^*) = p_0 \frac{\partial M}{\partial x^*} \quad \ldots(8)$$

$$H^*(T^*) = -p_0 \frac{\partial M}{\partial T^*} \quad \ldots(9)$$

where $H^*(T) = H(x^*(T), p_0, p(T), u^*(T), T)$

Comments

(1) For the constraints on u as given by (2) the equation (7) implies that along the optimal trajectory

$$\frac{\partial H}{\partial u_j} \begin{cases} \geq 0 & \text{if } u_j^* = \alpha_j \\ = 0 & \text{if } \alpha_j < u_j^* < \beta_j \\ \leq 0 & \text{if } u_j^* = \beta_j \end{cases} \quad j=1,\ldots,m$$

(2) If the terminal time is fixed, i.e.

$$T^* = T \sim \text{given} \quad \ldots(10)$$

then the condition (10) replaces (9)

(3) If some of the final states are fixed, say the first n_1, i.e.

$$x_i(T^*) = x_{if} \sim \text{given}, \quad i = 1,\ldots,n_1 \leq n \quad \ldots(11)$$

then the conditions (11) and

$$p_i(T^*) = p_0 \frac{\partial M}{\partial x_i^*} \qquad i = n_1+1,\ldots,n \qquad \ldots(12)$$

replace (8).

In order to extend theorem 1 to include the constraints of the form (3) let us define

$$\overline{M} = p_0 M + \nu' \psi \qquad \ldots(13)$$

where ν is a constant h-vector. Then the following theorem holds.

Theorem 2:

Let $u^*(t)$ be an optimal control and let $x^*(t)$ be the corresponding optimal trajectory. Then there exist multipliers $p_0 \geq 0$, $\nu_1,\ldots,\nu_h, p_1(t),\ldots,p_n(t)$, not vanishing simultaneously, such that (a), and (b) of theorem 1 hold and such that the following transversality conditions are fulfilled:

$$p(T^*) = \frac{\partial \overline{M}}{\partial x^*} \qquad \ldots(14)$$

$$H^*(T^*) = -\frac{\partial \overline{M}}{\partial T^*} \qquad \ldots(15)$$

We note that the comments immediately following Theorem 1 apply, mutatis mutandis, to Theorem 2.

We are now in a position to discuss some of the iterative procedures that have been developed for the solution of the optimal control problems. The basic idea behind all these procedures is to satisfy some of the necessary conditions outlined in Theorem 1 or Theorem 2 and then through iterative techniques satisfy the rest of them.

3.0 THE GRADIENT METHOD

For the sake of simplicity and clarity we shall first consider the problem where there are no constraints of the form (2) or (3).

We shall also assume that the final time T is given. From
Theorem 1, the necessary conditions which the optimal control, the
optimal trajectory and the associated multipliers satisfy,
are given by:

$$\dot{x} = \frac{\partial H}{\partial p}, \quad \dot{p} = -\frac{\partial H}{\partial x} = -p_0 \frac{\partial L}{\partial x} - \left(\frac{\partial f}{\partial x}\right)' p,$$

$$H(x^*, p_0, p, u^*, t) \leq H(x^*, p_0, p, u, t) \qquad \text{for all } u,$$

$$p(T) = p_0 \frac{\partial M}{\partial x}$$

where the vector $(p_0, p(t))$ never vanishes. It is easy to see that
$p_0 \neq 0$. In fact, if $p_0 = 0$ it follows from the last relation that
$p(T) = 0$ which contradicts the assumption that the vector $(p_0, p(t))$
never vanishes. Hence $p_0 > 0$. Moreover, it can be selected arbitrarily[*]
since the differential equations in p and the boundary conditions
$p(T) = p_0 \frac{\partial M}{\partial x}$ are respectively linear in (p_0, p) and $(p_0, p(T))$. By
choosing $p_0 = 1$, the necessary conditions become

$$\dot{x} = \frac{\partial H}{\partial p} = f(x, t, u) \qquad \ldots(16)$$

$$\dot{p} = -\frac{\partial H}{\partial x} = -\frac{\partial L}{\partial x} - \left(\frac{\partial f}{\partial x}\right)' p \qquad \ldots(17)$$

$$H(x^*, p, u^*t) \leq H(x^*, p, u, t) \quad \text{for all } u \qquad \ldots(18)$$

$$p(T) = \frac{\partial M}{\partial x} \qquad \ldots(19)$$

where $H = L + p'f$. The equations (17) are often called costate
equations.

In the gradient method (also called steepest descent method) a

---

[*] The arbitrariness of p_0 has been exploited in some computational
techniques, see [4] and [10].

nominal control $\hat{u}(t)$ is chosen and the state equations (16) with the given initial conditions are solved to determine the nominal trajectory. In general, this nominal trajectory is not the optimal trajectory, i.e. it does not yield the minimum value of the performance

$$J(u) = M(x(T)) + \int_{t_0}^{T} L(x,t,u)dt$$

A correction $\delta u(t)$ to the control $\hat{u}(t)$ is found such that $J(\hat{u}+\delta u) < J(\hat{u})$

Let $\hat{x} + \Delta x$ denote the solution of (16) with $u = \hat{u} + \delta u$ i.e.

$$(\hat{x}+\Delta x)^{\cdot} = f(\hat{x}+\Delta x, t, \hat{u}+\delta u), \quad \hat{x}(t_0) + \Delta x(t_0) = x_0$$

Since $\hat{x}$ satisfies

$$\dot{\hat{x}} = f(\hat{x},t,\hat{u}), \quad \hat{x}(t_0) = x_0$$

it follows that

$$\dot{\Delta x} = f(\hat{x}+\Delta x, t, \hat{u}+\delta u) - f(\hat{x},t,\hat{u}), \quad \Delta x(t_0) = 0 \quad \ldots\ldots(20)$$

Expanding the right-hand side of (20) into a Taylor series and retaining the first-order terms we find that $\delta x \simeq \Delta x$ satisfies

$$\delta \dot{x} = \left(\frac{\partial f}{\partial x}\right)\delta x + \left(\frac{\partial f}{\partial u}\right)\delta u \quad \delta x(t_0) = 0 \quad \ldots\ldots(21)$$

where the partial derivatives here and in the sequel will be understood to be evaluated along the nominal path.

Let us now calculate the first-order change in the performance:

$$\delta J \simeq J(\hat{u}+\delta u) - J(\hat{u}) \simeq M(\hat{x}(T)+\delta x(T)) - M(\hat{x}(T)) +$$

$$+ \int_{t_0}^{T} [L(\hat{x}+\delta x, t, \hat{u}+\delta u) - L(\hat{x},t,\hat{u})]dt$$

$$= \left(\frac{\partial M}{\partial x}\right)'\delta x(T) + \int_{t_0}^{T} \left[\left(\frac{\partial L}{\partial x}\right)'\delta x + \left(\frac{\partial L}{\partial u}\right)'\delta u\right]dt \quad \ldots\ldots(22)$$

Substituting the value of $\frac{\partial L}{\partial u}$ from $\frac{\partial H}{\partial u} = \frac{\partial L}{\partial u} + \left(\frac{\partial f}{\partial u}\right)' p$ into (22) we find that

$$\delta J = \left(\frac{\partial M}{\partial x}\right)' \delta x(T) + \int_{t_0}^{T} \left[\left(\frac{\partial L}{\partial x}\right)' \delta x + \left(\frac{\partial H}{\partial u}\right)' \delta u - p'\left(\frac{\partial f}{\partial u}\right)\delta u\right] dt \quad \ldots\ldots(23)$$

Assuming that (17) is satisfied and solving it for $\left(\frac{\partial L}{\partial x}\right)$ we find that $\left(\frac{\partial L}{\partial x}\right) = -\dot{p} - \left(\frac{\partial f}{\partial x}\right)' p$. Substituting the value of $\left(\frac{\partial L}{\partial x}\right)$ into (23) and utilizing (21) we get

$$\delta J = \left(\frac{\partial M}{\partial x}\right)' \delta x(T) + \int_{t_0}^{T} \left[-(\dot{p}'\delta x + p'\delta \dot{x}) + \left(\frac{\partial H}{\partial u}\right)' \delta u\right] dt$$

$$= \left(\frac{\partial M}{\partial x}\right)' \delta x(T) - p'\delta x(t) \Big|_{t_0}^{T} + \int_{t_0}^{T} \left(\frac{\partial H}{\partial u}\right)' \delta u \, dt$$

Assuming that (19) holds and remembering that $\delta x(t_0) = 0$, δJ is given by

$$\delta J = \int_{t_0}^{T} \left(\frac{\partial H}{\partial u}\right)' \delta u \, dt \quad \ldots\ldots(24)$$

We want to choose δu so that δJ is negative. In order to limit the correction $\delta u(t)$ so that the first-order relations obtained above remain valid we impose a constraint of the form

$$\int_{t_0}^{T} \delta u' \delta u \, dt = \varepsilon^2 \quad \ldots\ldots(25)$$

where ε is suitably chosen small number. We thus have a problem of determining δu which minimizes (24) subject to the constraint (25).

To this minimization problem the minimum principle can be applied. Applying the minimum principle or otherwise it is easy to check that

$$\delta u = - \frac{|\varepsilon|}{\left[\int_{t_0}^{T} \left(\frac{\partial H}{\partial u}\right)'\left(\frac{\partial H}{\partial u}\right) dt\right]^{1/2}} \left(\frac{\partial H}{\partial u}\right) \quad \ldots\ldots(26)$$

It is customary to write (26) in the form

$$\delta u = - k\left(\frac{\partial H}{\partial u}\right) \qquad k > 0 \quad \ldots\ldots(27)$$

Let us note here that in this method (16), (17) and (19) have been assumed to be satisfied.

The computer algorithm might proceed as follows:

(i) Pick a $\hat{u}(t)$ and solve the state equations (16) with the given initial conditions. Store $\hat{x}(t)$, $t_0 \leq t \leq T$.

(ii) Determine $p(T)$ by (19) and integrate (17) backwards from T to t_0. During the backward integration calculate $(\frac{\partial H}{\partial u})$.

(iii) Solve (16) forward with $u = \hat{u} - k(\frac{\partial H}{\partial u})$ for several values of k and determine the corresponding values of J.

(iv) Using a curve fitting technique, find the best value of k (the value that yields the least value of J) and use this value to determine the next estimate for u.

Return to step (i) and repeat. The process ends when the change in the performance tends to zero.

3.1 Min H Method and Conjugate Gradient Method

Closely related to the method of gradients are the Min H method [11] and a conjugate gradient method* as described in [12].

Min H Method

In this method a nominal control $\hat{u}(t)$ is chosen and just as in the gradient method the state and costate equations are integrated forwards and backwards respectively. However, instead of determining $(\frac{\partial H}{\partial u})$, H, considered as a function of u, is minimized. Let $\bar{u}$ denote the control that minimizes H. Then with $u = \hat{u} + k(\bar{u}-\hat{u})$ steps (iii) and (iv) of the gradient method are carried out.

Conjugate Gradient Method

Let u_i be the ith approximation to the optimal control and let $g_i \equiv (\frac{\partial H}{\partial u})_{u=u_i}$ be obtained by solving the state and costate

---

*See [13] - [15] for other conjugate gradient methods.

equations as in steps (i) and (ii) of the gradient method.

The computer algorithm proceeds as follows:

(i) select an arbitrary u_0

(ii) compute $s_0 = -g_0 = -\left(\frac{\partial H}{\partial u}\right)_{u=u_0}$

(iii) determine $\ell = \ell_i$ that minimizes $J(u_i + \ell s_i)$

(iv) set

$$u_{i+1} = u_i + \ell_i s_i$$

$$s_{i+1} = -g_{i+1} + m_i s_i$$

where

$$m_i = \frac{\|g_{i+1}\|^2}{\|g_i\|^2} \quad \text{and} \quad \|g\| = \int_{t_0}^{T} g'g \, dt.$$

It is shown in [12] that $J(u_{i+1}) < J(u_i)$, i.e. $J(u)$ is decreased at each step.

Remark

It has been assumed that the final time T is fixed. If, however, T is free then the condition

$$\left(\frac{\partial M}{\partial x}\right)' f(x(T),T,u(T)) + L(x(T),T,u(T)) + \left(\frac{\partial M}{\partial T}\right) = 0$$

is used as a stopping condition that determines the final time T. The above condition is obtained from (9) with $p(T)$ given by (8) and $p_0 = 1$.

3.2 Boundary Constraints

We shall now develop a steepest descent procedure [16] for

the computational solution of a control problem with constraints of the form (3). Let us assume that $p_0 \neq 0$. By choosing $p_0 = 1$, the necessary conditions for the control problem are given by (see Theorem 2)

$$\dot{x} = \frac{\partial H}{\partial x} \quad \ldots\ldots(28)$$

$$\dot{p} = -\left(\frac{\partial f}{\partial x}\right)' p - \frac{\partial L}{\partial x} \quad \ldots\ldots(29)$$

$$H(x^*, p, u^*, t) \leq H(x^*, p, u, t) \quad \text{for all } u \quad \ldots\ldots(30)$$

$$p(T) = \frac{\partial M}{\partial x} + \left(\frac{\partial \psi}{\partial x}\right)' \nu \quad \ldots\ldots(31)$$

$$H(T) = -\left(\frac{\partial M}{\partial T} + \nu' \frac{\partial \psi}{\partial T}\right) \quad \ldots\ldots(32)$$

where $H = L + p'f$. We shall use a component of ψ as a stopping condition that determines the final time T. Let it be the last component of ψ, i.e., ψ_h and let $\phi \equiv \psi_h$. Then from (31) and (32) we have

$$p(T) = \frac{\partial M}{\partial x} + \left(\frac{\partial \psi}{\partial x}\right)' \nu + \nu_h \frac{\partial \phi}{\partial x} \quad \ldots\ldots(33)$$

$$H(T) \equiv p'(T)f(T) + L(T) = -\left(\frac{\partial M}{\partial T} + \nu' \frac{\partial \psi}{\partial T} + \nu_h \frac{\partial \phi}{\partial T}\right) \quad \ldots\ldots(34)$$

where ν and ψ are now $(h-1)$-vectors and $f(T)$ and $L(T)$ denote the values of f and L at $t = T$ respectively. Substituting the value of $p(T)$ from (33) into (34) we get

$$\left[\left(\frac{\partial M}{\partial x}\right)' + \nu' \frac{\partial \psi}{\partial x} + \nu_h \left(\frac{\partial \phi}{\partial x}\right)'\right] f(T) + L(T) = -\left(\frac{\partial M}{\partial T} + \nu' \frac{\partial \psi}{\partial T} + \nu_h \frac{\partial \phi}{\partial T}\right) \quad \ldots\ldots(35)$$

From (35) we obtain

$$\boldsymbol{v}_h = -\frac{\dot{M} + \boldsymbol{v}'\dot{\boldsymbol{\psi}} + L(T)}{\dot{\phi}} \qquad (\dot{\phi} \neq 0)$$

where $\dot{\phi} = \frac{\partial \phi}{\partial T} + \left(\frac{\partial \phi}{\partial x}\right)' \dot{x}(T)$, $\dot{\boldsymbol{\psi}} = \frac{\partial \boldsymbol{\psi}}{\partial T} + \left(\frac{\partial \boldsymbol{\psi}}{\partial x}\right) \dot{x}(T)$ and $\dot{M} = \frac{\partial M}{\partial T} + \left(\frac{\partial M}{\partial x}\right)' \dot{x}(T)$. Substituting the value of $\boldsymbol{v}_h$ into (33) and rearranging terms we get

$$p(T) = (\tilde{M} + \tilde{\boldsymbol{\psi}}'\boldsymbol{v}) \qquad \ldots\ldots(36)$$

where

$$\tilde{M} = \left(\frac{\partial M}{\partial x} - \frac{\dot{M}+L(T)}{\dot{\phi}} \left(\frac{\partial \phi}{\partial x}\right)\right)$$

$$\tilde{\boldsymbol{\psi}} = \frac{\partial \boldsymbol{\psi}}{\partial x} - \frac{\dot{\boldsymbol{\psi}}}{\dot{\phi}} \left(\frac{\partial \phi}{\partial x}\right)'.$$

Let, as before, $\hat{u}$ denote the nominal control and $\delta u(t)$ denote a small perturbation about the nominal control. The change in the performance up to the first-order terms is given by

$$\delta J \simeq J(\hat{u}+\delta u) - J(\hat{u}) = \left(\frac{\partial M}{\partial x}\right)' \delta x(T) + \dot{M}\delta T + \int_{t_0}^{T} \left[\left(\frac{\partial L}{\partial x}\right)' \delta x + \left(\frac{\partial L}{\partial u}\right)' \delta u\right] dt$$
$$+ L(T)\delta T \qquad \ldots\ldots(37)$$

Evaluating the integral in (37) as in Section 3.0, we find that

$$\delta J = \left(\frac{\partial M}{\partial x}\right)' \delta x(T) + \dot{M}\delta T - p'(T)\delta x(T) + L(T)\delta T + \int_{t_0}^{T} \left(\frac{\partial H}{\partial u}\right)' \delta u dt \qquad \ldots\ldots(38)$$

Since $\phi(x(T),T) = 0$ is used as a stopping condition it follows that $\phi(x(T)+\Delta x(T), T+\delta T) = 0$ where $\Delta x(T) = \delta x(T)+f(T)\delta T$. Expanding $\phi(x(T)+\Delta x(T), T+\delta T) = 0$ into a Taylor series and retaining the first-order terms we have

$$\left(\frac{\partial \phi}{\partial x}\right)' \Delta x(T) + \frac{\partial \phi}{\partial T} \delta T = 0$$

or $\quad \left(\frac{\partial \phi}{\partial x}\right)' \delta x(T) + \dot{\phi} \delta T = 0$

or $\quad \delta T = -\dfrac{\left(\frac{\partial \phi}{\partial x}\right)' \delta x(T)}{\dot{\phi}}$

Substituting the value of δT into (38) we obtain

$$\delta J = [(\frac{\partial M}{\partial x})' - \frac{\dot{M}+L(T)}{\dot{\phi}} (\frac{\partial \phi}{\partial x})' - p'(T)]\delta x(T) + \int_{t_0}^{T} (\frac{\partial H}{\partial u})' \delta u\, dt$$

$$= (\tilde{M}' - p'(T))\delta x(T) + \int_{t_0}^{T} (\frac{\partial H}{\partial u})' \delta u\, dt$$

Let us denote the solution of (29) with $p(T) = \tilde{M}$ by $p^{(1)}(t)$ then

$$\delta J = \int_{t_0}^{T} (\frac{\partial H^{(1)}}{\partial u})' \delta u\, dt \qquad \ldots\ldots(39)$$

where $H^{(1)} = (p^{(1)})' f + L$.

Let $\delta \psi$ denote the change in ψ to the first-order of approximation. Then

$$\delta \psi = \frac{\partial \psi}{\partial x} \delta x(T) + \dot{\psi}\, \delta T = \tilde{\psi} \delta x(T) = \int_{t_0}^{T} \frac{d}{dt}(\theta' \delta x)\, dt$$

$$= \int_{t_0}^{T} (\dot{\theta}' \delta x + \theta' \delta \dot{x})\, dt = \int_{t_0}^{T} \theta'(\frac{\partial f}{\partial u})\delta u\, dt \qquad \ldots\ldots(40)$$

where θ satisfies $\dot{\theta} = -(\frac{\partial f}{\partial x})' \theta$ with $\theta(T) = \tilde{\psi}$.

For steepest descent we determine δu that minimizes δJ in equation (39) subject to the constraints

$$\int_{t_0}^{T} \delta u' \delta u\, dt = \varepsilon^2$$

$$\int_{t_0}^{T} \theta'(\frac{\partial f}{\partial u})\delta u\, dt = \delta \psi$$

for given values of ε and $\delta \psi$. The result of this minimization problem yields δu of the form

$$\delta u = -k \left(\frac{\partial H^{(1)}}{\partial u} - \left(\frac{\partial f}{\partial u}\right)' \theta M_1^{-1} L_1 \right) + \left(\frac{\partial f}{\partial u}\right)' \theta M_1^{-1} \delta \psi \quad ..(41)$$

where

$$M_1 = \int_{t_0}^T \theta' \left(\frac{\partial f}{\partial u}\right)\left(\frac{\partial f}{\partial u}\right)' \theta \, dt$$

$$L_1 = \int_{t_0}^T \theta' \left(\frac{\partial f}{\partial u}\right)\left(\frac{\partial H^{(1)}}{\partial u}\right) dt$$

$$k = \left[\frac{N_1 - L_1' M_1^{-1} L_1}{\varepsilon^2 - \delta \psi' M_1^{-1} \delta \psi} \right]^{-1/2}, \quad N_1 = \int_{t_0}^T \left(\frac{\partial H^{(1)}}{\partial u}\right)'\left(\frac{\partial H^{(1)}}{\partial u}\right) dt.$$

Substituting δu from (41) into (39) we find that

$$\delta J = -[(\varepsilon^2 - \delta \psi' M_1^{-1} \delta \psi)(N_1 - L_1' M_1^{-1} L_1)]^{1/2} + L_1' M_1^{-1} \delta \psi.$$

The computing procedure can now be summarized as follows:

(i) Pick a $\hat{u}(t)$ and solve the state equations with the given initial conditions until $\emptyset = 0$ and store the solution.

(ii) Determine $p^{(1)}(t)$ and $\theta(t)$ by integrating backwards from T to t_0 the equation (29) and $\dot{p} = -\left(\frac{\partial f}{\partial x}\right)' p$ respectively with appropriate boundary conditions.

(iii) Compute also L_1, M_1 and N_1.

(iv) Select a reasonable value of ε and set $\delta \psi = -a\psi$, $0 < a \leq 1$. Choose a and calculate $\varepsilon^2 - a^2 \psi' M_1^{-1} \psi$; if this is negative scale down a to make this quantity vanish. If the equality is positive, leave it as is.

(v) Determine δu by (41) and obtain a new nominal path by using $u = \hat{u} + \delta u.$*

Repeat the process until the terminal constraints $\psi = 0$ are satisfied and $N_1 - L_1' M_1^{-1} L_1$ tends to zero.

---

*If $T_{new} > T_{old}$, then the updated control u must be extrapolated. This may be done by setting $u(t) = u(T_{old})$, $T_{old} \leq t \leq T_{new}$.

We note that if the final time T is given then (32) is no more true and therefore p(T) given by (31) should be used in place of (36). Hence for the calculation of δu by the formula (41) $p^{(1)}(t)$ should be found as the solution of (29) with $p(T) = \frac{\partial M}{\partial x}$ and $\Theta(t)$ as the solution of $\dot{\Theta} = -(\frac{\partial f}{\partial x})'\Theta$ with $\Theta(T) = (\frac{\partial \psi}{\partial x})'$.

Another technique that has been widely used for handling problems with boundary constraints is the so-called penalty function technique [11]. In this scheme the constrained problem is reduced to an unconstrained problem by the introduction of a new functional

$$\hat{J} = M + \frac{1}{2} \sum_{j=1}^{h} K_j \psi_j^2 + \int_{t_0}^{T} L(x,t,u)dt$$

where $K_j, j=1,\ldots h$, are positive constants. Since the constraint violations add positive terms to $\hat{J}$, it is, intuitively, clear that for increasingly large positive values of K_j the unconstrained problem approximates to the constrained problem and hence the methods explained in the previous sections for the unconstrained problems can be applied.

In the application of the penalty function technique, numerically large values of K_j are needed. An efficient procedure for adjusting the values of K_j has been proposed by Kelley [11]. From (8) of Theorem 1, the transversality conditions for the above problem with $p_0 = 1$ are given by

$$p_i(T) = \frac{\partial M}{\partial x_i} + \sum_{j=1}^{h} K_j \psi_j \frac{\partial \psi_j}{\partial x_i} . \qquad \ldots\ldots(42)$$

Comparing (42) with (31) we see that $K_j \psi_j$ plays the role of ν_j and hence for adjusting K_j the following formula can be used

$$K_j = \left| \frac{\nu_j(\text{est.})}{\varepsilon_j} \right| \qquad j = 1,\ldots,h$$

where $\boldsymbol{\nu}_j$(est.) is the estimate of $\boldsymbol{\nu}_j$ and ε_j is the tolerance set on the constraint $\boldsymbol{\psi}_j = 0$. From (41) with $\delta\boldsymbol{\psi} = 0$ we get

$$\delta u = -k\left(\frac{\partial H}{\partial u}\right)$$

where $H = p'f+L$ with $p = p^{(1)} - \Theta M_1^{-1}L_1$ and therefore $\boldsymbol{\nu}$ can be estimated by

$$\boldsymbol{\nu}(\text{est.}) = -M_1^{-1}L_1$$

Note that Θ and $p^{(1)}$ that are needed in evaluating M_1 and L_1 are the solutions of $\dot{\Theta} = -\left(\frac{\partial f}{\partial x}\right)'\Theta$ with $\Theta(T) = \left(\frac{\partial \boldsymbol{\psi}}{\partial x}\right)'$ and the solution of (29) with $p(T) = \left(\frac{\partial M}{\partial x}\right)$ respectively.

Another procedure [11], perhaps less efficient than that discussed above, for adjustment of K_j values is the employment of the absolute value of violation/tolerance ratio as a factor to the current K_j values at the end of each step.

3.3 Problems With Control Constraints

Consider the problem posed in Section 3.0 and assume that there are now control constraints of the form (2). For the method of gradients we follow the same procedure as discussed in Section 3.0 except that for determining δu we solve the problem of minimizing:

$$\delta J = \int_{t_0}^{T} \left(\frac{\partial H}{\partial u}\right)'\delta u\, dt$$

Subject to

$$\int_{t_0}^{T} \delta u'\delta u\, dt = \varepsilon^2 \text{ and } \alpha_j - \hat{u}_j \leq \delta u_j \leq \beta_j - \hat{u}_j \qquad j=1,\ldots,m$$

Applying the results obtained by Berkovitz ([8], Theorem 2), it is easy to see that for vanishingly small step size ε

$$\delta u_j = - k(\frac{\partial H}{\partial u_j}), \quad k > 0$$

when $\hat{u}_j$ satisfies $\alpha_j < \hat{u}_j < \beta_j$;

$$\delta u_j = - k(\frac{\partial H}{\partial u_j}), \quad k > 0, \quad \text{if } (\frac{\partial H}{\partial u_j}) \leq 0$$

$$= 0 \quad \text{if } (\frac{\partial H}{\partial u_j}) > 0$$

when $\hat{u}_j = \alpha_j$;

$$\delta u_j = - k(\frac{\partial H}{\partial u_j}), \quad k > 0, \quad \text{if } (\frac{\partial H}{\partial u_j}) \geq 0$$

$$= 0 \quad \text{if } (\frac{\partial H}{\partial u_j}) < 0$$

when $\hat{u}_j = \beta_j$.

The above relations, in simple words, mean that the corrected controls $\hat{u}_j - k(\frac{\partial H}{\partial u_j})$ should be clipped off if they violate the control constraints. This gives the modification needed in the method of steepest descent to solve problems with constraints on the control variables.

For the Min H method, the modification needed is simple and H considered as a function of u is minimized subject to the constraints of the form (2). However, no comprehensive treatment[*] that modifies the conjugate gradient method is available at present.

The penalty function idea ([11],[17]) can also be applied to solve problems with control constraints by defining a suitable penalty function. Let the control constraints (2) be written in the form $g_i \geq 0$, $i = n+1,\ldots,s$. Define $x_i(t)$, $i = n+1,\ldots,s$ by

---

[*]An approach to adapt the clipping technique, mentioned in connection with the gradient method, is discussed in Ref. [14].

$$\dot{x}_i = g_i^2 \,, \quad g_i < 0$$
$$= 0 \,, \quad g_i \geq 0 \qquad x_i(t_0) = 0$$

Notice that the inequality constraints $g_i \geq 0$ are satisfied only when $x_i(T) = 0$. Thus the penalty terms corresponding to these terminal constraints can be added, as before, to the functional to form a new functional

$$\hat{J} = M + \sum_{i=n+1}^{s} K_i x_i(T) + \int_{t_0}^{T} L \, dt$$

to be minimized without inequality constraints $g_i \geq 0$. For the adjustment of K_i values the absolute value of violation/tolerance ratio can be employed as a factor to the current K_i values.

We shall now discuss another type of penalty function [17] which has the advantage of yielding solution that never violates the constraint. Let $g_j \geq 0$, $j = 1,\ldots,k$ denote the control constraints given in (2). Then the inequality constrained problem is converted to a problem without such constraints by defining a new functional

$$\hat{J} = M + \int_{t_0}^{T} L \, dt + r \sum_{j=1}^{k} \int_{t_0}^{T} \frac{dt}{g_j} \,, \quad r > 0.$$

It is shown in [17] that as $r \longrightarrow 0$ the minimum of $\hat{J}$ tends to the minimum J of the given problem, i.e. the problem with inequality constraints.

4.0 SUCCESSIVE SWEEP METHOD[18]

Let us consider a control problem in which the final time T is fixed and that there are no constraints on the control. Then, the necessary conditions from Theorem 2 are given by

$$\dot{x} = \frac{\partial H}{\partial p} \equiv H_p = f \quad \ldots(43)$$

$$\dot{p} = -\frac{\partial H}{\partial x} \equiv -H_x \quad \ldots(44)$$

$$H_u \equiv \frac{\partial H}{\partial u} = 0 \quad \ldots(45)$$

$$p(T) = \frac{\partial M}{\partial x} + \left(\frac{\partial \psi}{\partial x}\right)' \nu$$

$$\equiv M_x + \psi_x' \nu \quad \ldots(46)$$

To the above conditions we add the condition

$$\psi = 0 \quad \ldots(47)$$

and apply the Newton-Raphson* approximation to equations (43) - (47). Writing $x^{(n+1)}$ etc. for the (n+1)th iteration we obtain

$$\dot{x}^{(n+1)} - \dot{x}^{(n)} = H_{px}(x^{(n+1)} - x^{(n)}) + H_{pu}(u^{(n+1)} - u^{(n)}) - (\dot{x}^{(n)} - H_p) \quad \ldots(48)$$

$$\dot{p}^{(n+1)} - \dot{p}^{(n)} = -H_{xx}(x^{(n+1)} - x^{(n)}) - H_{xp}(p^{(n+1)} - p^{(n)}) - H_{xu}(u^{(n+1)} - u^{(n)}) - (\dot{p}^{(n)} + H_x) \quad \ldots(49)$$

$$H_u + H_{uu}(u^{(n+1)} - u^{(n)}) + H_{up}(p^{(n+1)} - p^{(n)}) + H_{ux}(x^{(n+1)} - x^{(n)}) = 0 \quad \ldots(50)$$

---

*Recall that in Newton-Raphson method for solving g = 0, the new point is obtained by setting the first order change equal to -g, i.e., $\left(\frac{\partial g}{\partial x}\right)\delta x = \delta g = -g$. A modification to this method is obtained by setting $\delta g = -ag$, $0 < a \leq 1$.

$$p^{(n+1)}(T) - p^{(n)}(T) = (M_{xx} + \nu^{(n)} \Psi_{xx})(x^{(n+1)}(T) - x^{(n)}(T)) + \Psi'_x (\nu^{(n+1)} - \nu^{(n)})$$
$$- (p^{(n)}(T) - M_x - \Psi'_x \nu^{(n)}) \quad \ldots (51)$$

$$\Psi_x (x^{(n+1)}(T) - x^{(n)}(T)) + \Psi^{(n)} = 0 \quad \ldots (52)$$

where all the partial derivatives are evaluated at the nth iteration, $g_{yz} = \frac{\partial}{\partial z}(\frac{\partial g}{\partial y})$ and $\nu^{(n)} \Psi_{xx} = \sum_{j=1}^{h} (\Psi_j)_{xx} \nu_j^{(n)}$.

As in the method of gradients we select a nominal control $\hat{u}$ and integrate (43) forward in time to obtain a nominal trajectory. Choose a value of ν and integrate (44) backward in time with initial conditions given by (46). In general (45) and (47) will not be satisfied. The perturbation around the nominal can be obtained from (48) - (52) by writing $x^{(n+1)} - x^{(n)} = \delta x$, $p^{(n+1)} - p^{(n)} = \delta p$, $u^{(n+1)} - u^{(n)} = \delta u$ and $\nu^{(n+1)} - \nu^{(n)} = \delta \nu$ and remembering that at the current iteration (43), (44) and (46) are satisfied. Thus we have

$$\delta \dot{x} = H_{px} \delta x + H_{pu} \delta u \quad \ldots (53)$$
$$\delta \dot{p} = -H_{xx} \delta x - H_{xp} \delta p - H_{xu} \delta u \quad \ldots (54)$$
$$a_1 H_u + H_{uu} \delta u + H_{up} \delta p + H_{ux} \delta x = 0 \quad \ldots (55)$$
$$\delta p(T) = (M_{xx} + \nu' \Psi_{xx}) \delta x(T) + \Psi'_x \delta \nu \quad \ldots (56)$$
$$a_2 \Psi = -\Psi_x \delta x(T) \quad \ldots (57)$$

where a_1 and $a_2 \in (0,1]$ (a modified Newton-Raphson approximation is used in equations (50) and (52)). Assuming that H_{uu} is non-singular, we have

$$\delta u = -H_{uu}^{-1}(a_1 H_u + H_{up} \delta p + H_{ux} \delta x) \quad \ldots (58)$$

Equation (58) gives the formula for determining the correction δu in terms of δx and δp that need to be determined.

A procedure to determine δx and δp will now be given. Substitute δu from (58) into (53) and (54) to obtain

$$\dot{\delta x} = A\delta x + B\delta p + v \qquad \ldots(59)$$

$$\dot{\delta p} = C\delta x - A'\delta p + w \qquad \ldots(60)$$

where $A = H_{px} - H_{pu}H_{uu}^{-1}H_{ux}$, $B = -H_{pu}H_{uu}^{-1}H_{up}$

$C = -H_{xx} + H_{xu}H_{uu}^{-1}H_{ux}$, $v = -a_1 H_{pu}H_{uu}^{-1}H_u$

and $w = a_1 H_{xu}H_{uu}^{-1}H_u$

Thus to obtain δx and δp we must solve (59) and (60) subject to $\delta x(t_0) = 0$, (56) and (57). To solve this problem introduce a Riccati transformation

$$\delta p = P\delta x + R\delta \nu + h \qquad \ldots(61)$$

$$-a_2 \Psi = R'\delta x + Q\delta \nu + g \qquad \ldots(62)$$

where P, R and Q are matrix functions, h and g are vector functions, and $\delta \nu$ is a constant vector. Differentiating (61) and (62) with respect to t we get

$$\dot{\delta p} = \dot{P}\delta x + P\dot{\delta x} + \dot{R}d\nu + \dot{h} \qquad \ldots(63)$$

$$0 = \dot{R}'\delta x + R'\dot{\delta x} + \dot{Q}\delta \nu + \dot{g} \qquad \ldots(64)$$

The elimination of δp from (59) by means of (61) yields

$$\dot{\delta x} = (A+BP)\delta x + BR\delta \nu + Bh + v \qquad \ldots(65)$$

Equating (60) and (63) and eliminating $\dot{\delta x}$ and δp by means of (65) and (61) we have

$$(C - A'P - PA - PBP - \dot{P})\delta x - [(A'+PB)R + \dot{R}]\delta \nu$$
$$-[(A'+PB)h + Pv - w + \dot{h}] = 0 \qquad \ldots(66)$$

Similarly, substituting (65) into (64) we get

$$[\dot{R} + R'(A+BP)]\delta x + (R'BR+\dot{Q})\delta\nu + R'(Bh+v)+\dot{g} = 0 \quad \ldots\ldots(67)$$

Regarding (66) and (67) as identities, valid for arbitrary values of δx and $\delta\nu$, we obtain the differential equations that P, R, Q, h and g satisfy. The boundary conditions are now obtained by comparing (56) and (57) with (61) and (62) at $t = T$. Thus

$$\dot{P} = -A'P - PA - PBP + C; \quad P(T) = M_{xx} + \nu'\psi_{xx} \quad \ldots\ldots(68)$$

$$\dot{R} = -(A' + PB) \quad ; \quad R(T) = \psi'_x \quad \ldots\ldots(69)$$

$$\dot{Q} = -R'BR \quad ; \quad Q(T) = 0 \quad \ldots\ldots(70)$$

$$\dot{h} = -(A'+PB)h - Pv + w \quad ; \quad h(T) = 0 \quad \ldots\ldots(71)$$

$$\dot{g} = R'(Bh + v) \quad ; \quad g(T) = 0 \quad \ldots\ldots(72)$$

The integration of (68) - (72) backward from T to t_0 determine P, R, Q, h and g (a 'backward sweep'). $\delta\nu$ can now be found from (62) at $t = t_0$.

$$\delta\nu = -Q^{-1}(t_0)(a_2\psi + g(t_0)) \quad \ldots\ldots(73)$$

(It is assumed that $Q^{-1}(t_0)$ exists)*. Having found $\delta\nu$, δx can be determined by (65) and hence δp and δu from (61) and (58) respectively.

We can also determine δu as a feedback law. Substituting the value of δp from (61) into (58) we have

$$\delta u = -H_{uu}^{-1}[a_1H_u + (H_{up}P + H_{ux})\delta x + H_{up}R\,\delta\nu + H_{up}h] \quad \ldots(74)$$

The value of $\delta\nu$ may be obtained from (73) or from (62) as

$$\delta\nu = -Q^{-1}[R'\delta x + g + a_2\psi] \quad \ldots(75)$$

---

*The question of existence of the inverse of Q and other related problems on the applicability of the sweep method may be found in [19].

Substituting the value of $\delta\nu$ from (75) into (74) we have

$$\delta u = -H_{uu}^{-1}[(H_{ux}+H_{up}(P-RQ^{-1}R'))\delta x - a_2 H_{up}RQ^{-1}\psi + a_1 H_u + H_{up}(h-RQ^{-1}g)]$$

$$\dots(76)$$

To start the next iteration we solve (43) forward (a 'forward sweep') with

$$u = \hat{u} + \delta u \text{ and } \delta x = x - \hat{x}$$

where δu is determined from (74) with $\delta\nu$ given by (73) or δu is given by (76). The new $p(t)$ is determined by integrating (44) backward with boundary conditions given by (46) in which ν is corrected by the formula (73). The process ends when $H_u = 0$ and $\psi = 0$.

4.1 Final Time Given Implicitly

The procedure for the fixed final time case can be easily extended to the case of final time that is given implicitly by the introduction of some additional equations. We shall not discuss this extension here and simply refer to McReynolds and Bryson [18] for detail. However, we mention an alternative procedure in which a sequence of fixed terminal time problems is solved.

From (15) (see Theorem 2), the additional necessary condition for free terminal time is

$$\Omega \equiv (\frac{\partial \overline{M}}{\partial x})'f(T)+L(T)+(\frac{\partial \overline{M}}{\partial T}) = 0 \quad \dots(77)$$

Pick a suitable final time T_0 and solve the implicitly given terminal time problem as a fixed terminal time problem by setting $T = T_0$. For the terminal time T_0 determine Ω.

Normally $\Omega \neq 0$. Repeat this for another neighbouring value $T = T_1$. The final time T can now be automatically adjusted by the application of a discrete modified Newton-Raphson method:

$$T_{k+2} = T_{k+1} + \frac{T_{k+1} - T_k}{\Omega(T_{k+1}) - \Omega(T_k)} (- a \Omega(T_{k+1})), \quad k \geq 0$$

where $0 < a \leq 1$.

Finally, we mention that the problem with constraints on the control variables is solved by the penalty function technique.

5.0 SECOND-VARIATION METHOD

The second-variation method ([20]-[22]) is a direct extension of the steepest descent method which is based on first-order theory. Here, in this method, second-order effects are included in the expansion of ΔJ and an auxiliary minimization problem is posed. This minimization problem results in a two-point boundary value problem which is the same as the one discussed in the successive sweep method. Since the two methods lead to the same auxiliary two-point boundary value problem, the method will be illustrated by means of a problem with fixed terminal time and no constraints of the form (2) or (3).

As in the method of gradients, select a nominal control $\hat{u}$ and obtain a nominal trajectory. Let $\hat{x}+\Delta x$ denote the solution of the state equations with $u = \hat{u}+\delta u$:

$$(\hat{x}+\Delta x)^{\cdot} = f(\hat{x}+\Delta x, t, \hat{u}+\delta u), \quad \hat{x}(t_0) + \Delta x(t_0) = x_0$$

Then

$$\Delta \dot{x} = f(\hat{x}+\Delta x, t, \hat{u}+\delta u) - f(\hat{x}, t, \hat{u}), \quad \Delta x(t_0) = 0$$

The change in the performance is given by

$$\Delta J = J(\hat{u}+\delta u) - J(\hat{u}) = M(\hat{x}(T)+\Delta x(T))-M(\hat{x}(T))$$
$$+ \int_{t_0}^{T} [L(\hat{x}+\Delta x,t,\hat{u}+\delta u)-L(\hat{x},t,\hat{u})]dt$$

Since $H = L+p'f$ we have

$$\Delta J = M(\hat{x}(T)+\Delta x(T))-M(\hat{x}(T)) + \int_{t_0}^{T} [H(\hat{x}+\Delta x,p,\hat{u}+\delta u,t)-H(\hat{x},p,\hat{u},t)$$
$$- p'(f(\hat{x}+\Delta x,t,\hat{u}+\delta u) - f(\hat{x},t,\hat{u})]dt$$

Expanding ΔJ into a Taylor series and retaining the second-order terms in Δx and δu we find

$$\Delta J \simeq M_x' \Delta x(T) + \tfrac{1}{2} \Delta x(T)'M_{xx}\Delta x(T)$$
$$+ \int_{t_0}^{T} [(H_x' \Delta x - p'\Delta \dot{x}) + H_u' \delta u + \tfrac{1}{2} \Delta x' H_{xx}\Delta x$$
$$+ \tfrac{1}{2} \Delta x' H_{xu}\delta u + \tfrac{1}{2} \delta u' H_{ux}\Delta x + \tfrac{1}{2} \delta u' H_{uu}\delta u]dt$$

Since $\dot{p} = -H_x$ it follows that

$$\int_{t_0}^{T} (H_x' \Delta x - p'\Delta \dot{x})dt = -\int_{t_0}^{T} \tfrac{d}{dt}(p'\Delta x)dt$$
$$= - (p'\Delta x) \Big|_{t_0}^{T} = -p' \Delta x(T) = -M_x' \Delta x(T)$$

Hence ΔJ, correct to second-order terms in control variations, can be written as

$$\Delta J = \tfrac{1}{2} \delta x(T)' M_{xx}\delta x(T) + \int_{t_0}^{T} [H_u'\delta u + \tfrac{1}{2} W(t)]dt \qquad \ldots\ldots(78)$$

where

$$W(t) = \delta x' H_{xx}\delta x + 2\delta x' H_{xu}\delta u + \delta u' H_{uu}\delta u$$

and δx satisfies

$$\delta \dot{x} = \frac{\partial f}{\partial x} \delta x + \frac{\partial f}{\partial u} \delta u \qquad \delta x(t_0) = 0 \qquad \ldots(79)$$

In order to determine δu, the following auxiliary optimal problem* is posed: Determine δu that minimizes (78) subject to (79). The necessary conditions given by the minimum principle are

$$\delta \dot{x} = f_x \delta x + f_u \delta u, \qquad \delta x(t_0) = 0 \qquad \ldots(80)$$

$$\dot{\lambda} = -f_x' \lambda - H_{xx} \delta x - H_{xu} \delta u, \lambda(T) = M_{xx} \delta x(T) \ldots(81)$$

$$H_u + H_{ux} \delta x + f_u' \lambda + H_{uu} \delta u = 0 \qquad \ldots(82)$$

Remembering that in this problem there are no terminal constraints and identifying λ with δp we see that the above equations are the same as those obtained in (53) - (56) ($a_1 = 1$). In order to determine δu from equations (80) - (82) we solve (82) for δu in terms of δx, $\delta \lambda$ and H_u:

$$\delta u = - H_{uu}^{-1} [H_u + H_{ux} \delta x + f_u' \lambda] \qquad \ldots(83)$$

The substitution of δu from (83) into (80) and (81) leads to a two-point boundary value problem

$$\delta \dot{x} = A \delta x + B \lambda + v \qquad \ldots(84)$$
$$\dot{\lambda} = C \delta x - A' \lambda + w \qquad \ldots(85)$$
$$\delta x(t_0) = 0, \quad \lambda(T) = M_{xx} \delta x(T) \qquad \ldots(86)$$

which can be solved in several ways [23]. (See Sec. 4 for the

---
*To limit the step size a restriction of the form $\int_{t_0}^{T} \delta u' \delta u \, dt = \varepsilon^2$ may also be imposed.

definitions of A,B,C, v and w.)

Let us apply the method of adjoint equations to the above formulated two-point boundary value problem. Write (84) - (86) in the form

$$\dot{z} = Dz + \hat{f} \qquad \ldots(87)$$
$$\hat{B}z(t_0) = 0, \quad \hat{C}z(T) = 0$$

where $z = \begin{pmatrix} \delta x \\ \lambda \end{pmatrix}$, $\hat{f} = \begin{pmatrix} v \\ w \end{pmatrix}$, $\hat{B} = (I\ 0)$, I, nxn unit matrix, $\hat{C} = (-M_{xx}\ I)$ and $D = \begin{pmatrix} A & B \\ C & -A' \end{pmatrix}$

Let the system adjoint to (87) be

$$\dot{y} = -D'y \qquad \ldots(88)$$

Then

$$y'(T)z(T) - y'(t_0)z(t_0) = \int_{t_0}^{T} y'\hat{f}\ dt \qquad \ldots(89)$$

Denote by $\hat{C}_j$, the jth row of $\hat{C}$ and the solution of (88) (solved backwards from T to t_0) with $\hat{C}_j'$ as the initial conditions by $y^{(j)}$. Then from (89) it follows that

$$\hat{C}_j z(T) - y'^{(j)}(t_0)z(t_0) = \int_{t_0}^{T} y'^{(j)}\hat{f}\ dt \qquad \ldots(90)$$

Noting that $\hat{C}z(T) = 0$ we obtain

$$y'^{(j)}(t_0)z(t_0) = \int_{T}^{t_0} y'^{(j)}\hat{f}\ dt \quad j = 1,\ldots,n \qquad \ldots(91)$$

from (90). The system of equations (91) together with $Bz(t_0) = 0$ constitute 2n linear equations to determine $z(t_0)$. Having determined $z(t_0)$ integrate (87) forward from t_0 to T to determine $z = \begin{pmatrix} \delta x \\ \lambda \end{pmatrix}$ and hence δu by (83).

We mention that the application of the Riccati transformation method to the linear two-point boundary value problem yields the algorithm obtained in Section 4.

6.0 SHOOTING METHODS

In the shooting methods (variously called perturbation methods, initial-costate search methods, boundary value iteration methods, indirect methods), u is eliminated from equations (5) by means of (7) and a two-point boundary value problem of the form

$$\dot{x} = f_1(x,p,t) \quad \ldots\ldots(92)$$
$$\dot{p} = f_2(x,p,t) \quad \ldots\ldots(93)$$
$$x(t_0) = x_0, \quad \psi(x(T),T) = 0$$
$$p(T) - \frac{\partial \overline{M}}{\partial x} = 0, \quad p'(T)f_1(T)+L(T)+ \frac{\partial \overline{M}}{\partial T} = 0 \quad \ldots(94)$$

is obtained. This two-point boundary value problem is then solved by iteration on the unknowns $p(t_0)$, ν and T in the following way. The equations (92) and (93) are integrated forward from t_0 to T (nominal) with $x(t_0) = x_0$ and a nominal value of $p(t_0)$. The (n+h+1) equations, $\psi = 0$ and (94), which depend on $p(t_0)$, ν and T (since p(T) and x(T) depend on $p(t_0)$ and T) will not be satisfied in general. Through various iterative procedures ([24]-[26]) $p(t_0)$, ν and T are found that satisfy the (n+h+1) nonlinear equations, $\psi = 0$ and (94).

Here we shall explain only one method--the Newton-Raphson method. The application of the other methods can be carried out similarly and will be briefly mentioned.

6.1 Newton-Raphson Method ([27]-[29])

Let us consider a control problem with no constraints on the control variables*. The necessary conditions given by

---

*For the problem with control constraints it is assumed that they are represented either by the introduction of new states (see [28]) or through the penalty function technique.

the minimum principle are

$$\dot{x} = H_p \qquad x(t_0) = x_0 \quad \ldots\ldots(95)$$

$$\dot{p} = -H_x \quad \ldots\ldots(96)$$

$$H_u = 0 \quad \ldots\ldots(97)$$

$$\zeta \equiv p(T) - \frac{\partial \overline{M}}{\partial x} = 0 \quad \ldots\ldots(98)$$

$$\Omega \equiv H(T) + \frac{\partial \overline{M}}{\partial T} = 0 \quad \ldots\ldots(99)$$

$$\psi = 0 \quad \ldots\ldots(100)$$

where $H = p'f + L$ and $\overline{M} = M + \nu' \psi$.

Let us pick a $p(t_0)$ and integrate forward the equations (95) and (96) with u determined from (97) until some terminal time T. The equations (98) - (100), in general, will not be satisfied with $p(T)$ and $x(T)$ so obtained and a guessed value for ν. Let us denote by w the (n+h+1)-vector $(p(t_0), T, \nu)$ and by G the vector (ζ, ψ, Ω). Then the problem is to determine w such that $G(w) = 0$. The recurrence relation for the Newton-Raphson method* is given by

$$w^{(k+1)} = w^{(k)} - (J^{(k)})^{-1} G(w^{(k)}) \qquad k = 1, 2, \ldots \quad \ldots(101)$$

where

$$J^{(k)} = \begin{pmatrix} \frac{\partial G_1}{\partial w_1} & \cdots & \frac{\partial G_1}{\partial w_{n+h+1}} \\ \vdots & & \\ \frac{\partial G_{n+h+1}}{\partial w_1} & \cdots & \frac{\partial G_{n+h+1}}{\partial w_{n+h+1}} \end{pmatrix}_{w^{(k)}}$$

In order to use (101) for updating w, we require $(J^{(k)})^{-1}$. To get this we first determine $J^{(k)}$. Of the several methods that determine

---

*A modified Newton-Raphson method may be applied in place of Newton-Raphson method.

$J^{(k)}$ numerically, perhaps the simplest method is the employment of a finite difference scheme such as

$$\frac{\partial G_i}{\partial w_j} = \frac{G_i(w_1,\ldots,w_j+\varepsilon,\ldots) - G_i(w_1,\ldots,w_j,\ldots)}{\varepsilon} \quad (\varepsilon > 0,\text{ sufficiently small})$$

or

$$\frac{\partial G_i}{\partial w_j} = \frac{G_i(w_1,\ldots,w_j+\varepsilon,\ldots) - G_i(w_1,\ldots,w_j-\varepsilon,\ldots)}{2\varepsilon}$$

The other methods of determining $J^{(k)}$ involve the determination of missing initial conditions in the solution of a linear two-point boundary value problem which can be formulated in the following way. Applying Newton-Raphson approximation to (95) - (97) and noting that (95) - (97) are satisfied at each iteration we obtain (see (53)-(55))

$$\delta\dot{x} = H_{px}\delta x + H_{pu}\delta u \qquad \ldots(102)$$

$$\delta\dot{p} = -H_{xx}\delta x - H_{xp}\delta p - H_{xu}\delta u \qquad \ldots(103)$$

$$H_{ux}\delta x + H_{up}\delta p + H_{uu}\delta u = 0 \qquad \ldots(104)$$

Notice that (102) and (103) are the linear variational equation to (95) and (96) respectively. Eliminate δu from (102) and (103) by means of (104) to obtain

$$\delta\dot{x} = A\delta x + B\delta p \qquad \delta x(t_0) = 0 \qquad \ldots(105)$$

$$\delta\dot{p} = C\delta x - A'\delta p \qquad \ldots(106)$$

where

$$A = H_{px} - H_{pu}H_{uu}^{-1}H_{ux} \qquad \ldots(107)$$

$$B = -H_{pu}H_{uu}^{-1}H_{up} \qquad \ldots(108)$$

$$C = -H_{xx} + H_{xu}H_{uu}^{-1}H_{ux} \qquad \ldots(109)$$

The first-order change in ζ, ψ and Ω can be written as

$$\delta\zeta = \delta p(T) - H_x(T)\delta T - \frac{\partial^2 \bar{M}}{\partial x^2}(\delta x(T)+H_p(T)\delta T)$$

$$- \frac{\partial^2 \bar{M}}{\partial \nu \partial x}\delta\nu - \frac{\partial^2 \bar{M}}{\partial T \partial x}\delta T$$

$$= \delta p(T) - \bar{M}_{xx}\delta x(T) - \psi'_x \delta\nu - (H_x(T)+\bar{M}_{xx}H_p+\bar{M}_{xT})\delta T$$

$$\delta\psi = \psi_x(\delta x(T) + \dot{x}(T)\delta T) + \psi_T \delta T$$

$$= \psi_x \delta x(T) + \dot{\psi}\delta T$$

$$\delta\Omega = H'_x(T)(\delta x(T)+H_p(T)\delta T) + H'_p(T)(\delta p(T)-H_x(T)\delta T)$$

$$+ H'_u(T)(\delta u(T)+\dot{u}(T)\delta T) + H_T(T)\delta T + \left(\frac{\partial^2 \bar{M}}{\partial x \partial T}\right)'(\delta x(T)+H_p(T)\delta T) +$$

$$+ \frac{\partial^2 \bar{M}}{\partial \nu \partial T}\delta\nu + \bar{M}_{TT}\delta T$$

$$= (H'_x(T)+\bar{M}'_{Tx})\delta x(T) + \psi'_T \delta\nu + (H_T(T)+\bar{M}_{Tx}H_p(T)+\bar{M}_{TT})\delta T$$

$$+ f'(T)\delta p(T). \quad \text{(Note that } H_u = 0 \text{ at each iteration.)}$$

Let us write

$$\bar{M}_{xx} = \alpha, \quad \psi'_x = \beta, \quad (H_x(T)+\bar{M}_{xx}H_p+\bar{M}_{xT}) = \gamma, \quad \dot{\psi} = \eta$$

$$H_x(T) + \bar{M}_{Tx} = \hat{\gamma}, \quad \psi_T = \hat{\eta}, \quad \mu = H_T(T) + \bar{M}_{Tx}H_p(T) + \bar{M}_{TT}.$$

Then

$$\delta\zeta = \delta p(T) - \alpha\delta x(T) - \beta\delta\nu - \gamma\delta T \qquad \ldots(110)$$

$$\delta\psi = \beta'\delta x(T) + \eta\delta T \qquad \ldots(111)$$

$$\delta\Omega = \hat{\gamma}'\delta x(T) + \hat{\eta}'\delta\nu + \mu\delta T + f'(T)\delta p(T). \qquad \ldots(112)$$

The equations (105), (106) and (110) - (112) with $\delta\zeta$, $\delta\psi$ and $\delta\Omega$ prescribed, form a linear two-point boundary value problem.

The method of complementary functions [27] or the method of adjoint equations [28] may be employed to determine $J^{(k)}$. Let us work out in detail the application of the complementary functions method for the determination of $J^{(k)}$. Denoting the transition matrix of (105) and (106) by $\Phi(t,t_0)$, the solution of (105) - (106) with $\delta x(t_0) = 0$ can be written as

$$\begin{pmatrix} \delta x \\ \delta p \end{pmatrix} = \Phi(t,t_0) \begin{pmatrix} \delta x(t_0) \\ \delta p(t_0) \end{pmatrix}$$

$$= \begin{pmatrix} \Phi_{11}(t,t_0) & \Phi_{12}(t,t_0) \\ \Phi_{21}(t,t_0) & \Phi_{22}(t,t_0) \end{pmatrix} \begin{pmatrix} 0 \\ \delta p(t_0) \end{pmatrix}$$

Let $\Phi_{12}(T,t_0)$ and $\Phi_{22}(T,t_0)$ respectively be denoted by Φ_{12T} and Φ_{22T}, then

$$\begin{pmatrix} \delta x(T) \\ \delta p(T) \end{pmatrix} = \begin{pmatrix} \Phi_{12T} \\ \Phi_{22T} \end{pmatrix} \delta p(t_0) \quad \ldots\ldots(113)$$

Substituting the values of $\delta x(T)$ and $\delta p(T)$ from (113) into (110) - (112) we get

$$\delta G = N \begin{pmatrix} \delta p(t_0) \\ \delta \nu \\ \delta T \end{pmatrix}$$

where

$$N = \begin{pmatrix} \Phi_{22T} - \alpha\Phi_{12T} & -\beta & -\gamma \\ \beta'\Phi_{12T} & 0 & \eta \\ f'(T)\Phi_{22T} + \hat{\gamma}'\Phi_{12T} & \hat{\eta}' & \mu \end{pmatrix}, \text{ evaluated at } w^{(k)},$$

gives $J^{(k)}$, $k = 1,2,\ldots$. Having determined $J^{(k)}$, the next value of the unknown vector w is determined by (101) and the process is repeated until ζ, Ω and ψ are all equal to zero.

Let us discuss now the application of the Riccati transformation method for the determination of δw. For this we substitute the value of $\delta p(T)$ from (110) into (112) and we get

$$\delta\Omega = \hat{\gamma}'\delta x(T) + \hat{\eta}'\delta\nu + \mu\delta T + f'(T)[\delta\zeta + \alpha\delta x(T) + \beta\delta\nu + \gamma\delta T]$$

$$= (\hat{\gamma}' + f'(T)\alpha)\delta x(T) + (\hat{\eta}' + f'(T)\beta)\delta\nu + (\mu + f'(T)\gamma)\delta T + f'(T)\delta\zeta$$

$$= \gamma'\delta x(T) + \eta'\delta\nu + (\mu + f'(T)\gamma)\delta T + f'(T)\delta\zeta \qquad \ldots\ldots(114)$$

Recalling that in the Newton-Raphson method* $\delta\zeta = -\zeta = -(p(T) - \overline{M}_x)$, $\delta\psi = -\psi$ and $\delta\Omega = -\Omega = -(H(T) + \overline{M}_T)$ we obtain from (110), (111) and (114)

$$\delta p(T) = \alpha\delta x(T) + \beta\delta\nu + \gamma\delta T - (p(T) - \overline{M}_x) \qquad \ldots\ldots(115)$$

$$-\psi = \beta'\delta x(T) + \eta\delta T \qquad \ldots\ldots(116)$$

$$-(L(T) + \dot{\overline{M}}) = \hat{\gamma}'\delta x(T) + \eta'\delta\nu + (\mu + f'(T)\gamma)\delta T \qquad \ldots\ldots(117)$$

where $L(T) = H(T) - p'(T)f(T)$ and $\dot{\overline{M}} = \overline{M}_T + \overline{M}'_x \dot{x}(T)$.

The equations (115) - (117) suggest the following Riccati transformation

$$\delta p = P\delta x + R\delta\nu + h_1\delta T + h \qquad \ldots\ldots(118)$$

$$-\psi = R'\delta x + Q\delta\nu + g_1\delta T + g \qquad \ldots\ldots(119)$$

$$-(L(T) + \dot{\overline{M}}) = h'_1\delta x + g'_1\delta\nu + \lambda_1\delta T + \lambda \qquad \ldots\ldots(120)$$

where $\delta\nu$, δT, ψ and $(L+\dot{\overline{M}})$ are constants and the others are vector or

---

*A modified Newton-Raphson method may be applied in place of the Newton-Raphson method by setting $\delta\zeta = -a_1\zeta$ $\delta\psi = -a_2\psi$, $\delta\Omega = -a_3\Omega$ where $0 < a_j \leq 1$, $j = 1,2,3$.

matrix functions of time. Following the procedure discussed in Section 4, we find that the differential equations and the boundary conditions that $P, R, \ldots, \lambda$ must satisfy are given by

$$\dot{P} = -PA - A'P - PBP + C \qquad P(T) = \alpha \qquad \ldots\ldots(121)$$
$$\dot{R} = -(A'+PB)R \qquad R(T) = \beta \qquad \ldots\ldots(122)$$
$$\dot{h}_1 = -(A'+PB)h_1 \qquad h_1(T) = \gamma \qquad \ldots\ldots(123)$$
$$\dot{h} = -(A'+PB)h \qquad h(T) = -(p(T)-\vec{M}_x) \qquad \ldots\ldots(124)$$
$$\dot{Q} = -R'BR \qquad Q(T) = 0 \qquad \ldots\ldots(125)$$
$$\dot{g}_1 = -R'Bh_1 \qquad g_1(T) = \eta \qquad \ldots\ldots(126)$$
$$\dot{g} = -R'Bh \qquad g(T) = 0 \qquad \ldots\ldots(127)$$
$$\dot{\lambda}_1 = -h_1'Bh_1 \qquad \lambda_1(T) = (\mu+f'(T)\gamma) \qquad \ldots\ldots(128)$$
$$\dot{\lambda} = h_1'Bh \qquad \lambda(T) = 0 \qquad \ldots\ldots(129)$$

We note that R, h_1 and h satisfy the same differential equation, namely $\dot{z} = -(A'+PB)z$.

The equations (121) - (129) are integrated backward from T to t_0 and $(\delta \boldsymbol{\nu}, \delta T)$ and $\delta p(t_0)$ are determined respectively from the equations

$$\begin{pmatrix} Q(t_0) & g_1(t_0) \\ g_1'(t_0) & \lambda_1(t_0) \end{pmatrix} \begin{pmatrix} \delta \boldsymbol{\nu} \\ \delta T \end{pmatrix} = - \begin{pmatrix} \boldsymbol{\psi} + g(t_0) \\ L(T)+\dot{\vec{M}}+\lambda(t_0) \end{pmatrix}$$

and

$$\delta p(t_0) = R(t_0)\delta \boldsymbol{\nu} + h_1(t_0)\delta T + h(t_0)$$

Having determined $\delta p(t_0)$, equations (95) and (96) (u determined from (97)) are integrated forward with the improved guess of $p(t_0)$ and the process is repeated until ζ, Ω and $\boldsymbol{\psi}$ are equal to zero.

We mention that for the fixed terminal time problem the equations (123), (126) and (128) and (129) are not needed and that the variable terminal time problem may be solved as a sequence of fixed terminal time problems (see Section 4.1).

6.2 Minimizing Methods ([30], [31])

In these methods the problem of determining w which satisfies $G(w) = 0$ is replaced by the problem of minimizing a scalar function $E(w)$, where, for example,

$$E(w) = \tfrac{1}{2}(G(w))'G(w) \quad \ldots(130)$$

The function $E(w)$ can be minimized in several ways ([32]-[35]). Let us discuss briefly the application of the steepest descent method. For simplicity assume that T is given explicitly. Guess a value $\hat{w}$ of w and integrate (95) and (96) forward with u determined from (97). The substitution of the values of $p(T)$, $\pmb{\psi}$ and $\bar{M}_x$ will not, in general, make $E(w)$ vanish where

$$E(w) \equiv E(p(t_0), \pmb{\nu}) = \tfrac{1}{2}(p(T)-\bar{M}_x)'(p(T)-\bar{M}_x) + \tfrac{1}{2}\pmb{\psi}'\pmb{\psi} \quad \ldots(131)$$

The correction δw in the steepest descent method is given by

$$\delta w = -k\left(\tfrac{\partial E}{\partial w}\right), \quad k > 0 \quad \ldots(132)$$

where, as before, the partial derivatives are evaluated at the nominal point $\hat{w}$.

In order to apply (132) we require the partial derivatives $\left(\tfrac{\partial E}{\partial w}\right)$. This can be determined either by the finite difference methods or by the complementary functions method or by the adjoint equations method. Let us determine the partial derivatives using the method of complementary functions.

The first-order variation in E is given by

$$\delta E = (p(T)-\overline{M}_x)'(\delta p(T)-\delta \overline{M}_x) + \psi'\delta\psi$$

$$= (p(T)-\overline{M}_x)'(\delta p(T)-\overline{M}_{xx}\delta x(T)-\overline{M}_{x\nu}\delta\nu) + \psi'\psi_x\delta x(T) \quad \ldots (133)$$

where $\delta x(T)$ and $\delta p(T)$ are given by (113). Substituting the values of $\delta x(T)$ and $\delta p(T)$ from (113) into (133) we get

$$\delta E = (p(T)-\overline{M}x)'(\Phi_{22T}\delta p(t_0)-\overline{M}_{xx}\Phi_{12T}\delta p(t_0)-\overline{M}_{x\nu}\delta\nu)$$

$$+ \psi'\psi_x\Phi_{12T}\delta p(t_0)$$

$$= \left((p(T)-\overline{M}_x)'(\Phi_{22T}-\overline{M}_{xx}\Phi_{12T}) + \psi'\psi_x\Phi_{12T}, (p(T)-\overline{M}_x)'(-\psi'_x)\right) \cdot$$

$$\begin{pmatrix} \delta p(t_0) \\ \delta\nu \end{pmatrix} \quad \ldots (134)$$

Hence

$$\left(\frac{\partial E}{\partial w}\right) = \begin{pmatrix} (\Phi_{22T}-\overline{M}_{xx}\Phi_{12T})'(p(T)-\overline{M}_x)+\Phi'_{12T}\psi'_x\psi \\ -\psi_x(p(T)-\overline{M}_x) \end{pmatrix}$$

Having determined $\left(\frac{\partial E}{\partial w}\right)$, the differential equations are integrated forward with $p(t_0) = \hat{p}(t_0) - k\left(\frac{\partial E}{\partial p(t_0)}\right)$ and $E(\hat{w} - k(\frac{\partial E}{\partial w}))$ is calculated. Using a one-dimensional minimization technique k is found that yields the minimum value of $E(\hat{w} - k(\frac{\partial E}{\partial w}))$. With this value of k the correction δw is determined from (132) and the process is repeated until $E = 0$.

7.0 THE GENERALIZED NEWTON-RAPHSON METHOD

Let us consider a control problem with no constraints on the control variables. Then the necessary conditions given by the minimum principle are

$$\dot{x} - H_p = 0, \quad \dot{p} + H_x = 0, \quad H_u = 0, \quad \zeta \equiv p(T) - \frac{\partial \overline{M}}{\partial x} = 0$$

$$\Omega \equiv H(T) + \frac{\partial \overline{M}}{\partial T} = 0, \quad \psi = 0$$

where $H = p'f+L$ and $\bar{M} = M+\nu'\psi$. Applying the Newton-Raphson* approximation to the above equations and denoting the points at the (n+1)th and nth iteration by $(\bar{x},\bar{p},\bar{u},\bar{\nu},\bar{T})$ and (x,p,u,ν,T) respectively we get

$$\dot{\bar{x}} - \dot{x} - H_{px}(\bar{x}-x) - H_{pu}(\bar{u}-u) = -(\dot{x}-H_p) \quad \ldots(136)$$

$$\dot{\bar{p}} - \dot{p} + H_{xx}(\bar{x}-x) + H_{xp}(\bar{p}-p) + H_{xu}(\bar{u}-u) = -(\dot{p}+H_x) \quad \ldots(137)$$

$$H_{ux}(\bar{x}-x) + H_{up}(\bar{p}-p) + H_{uu}(\bar{u}-u) = -H_u \quad \ldots(138)$$

$$\bar{p}(T) - p(T) + \dot{p}(T)(\bar{T}-T) - \bar{M}_{xx}(\bar{x}(T)-x(T)+\dot{x}(T)(\bar{T}-T)) - \bar{M}_{x\nu}(\bar{\nu}-\nu)$$
$$- \bar{M}_{xT}(\bar{T}-T) = -(p(T)-\bar{M}_x) \quad \ldots(139)$$

$$(H_x(T)+\bar{M}_{Tx})'(\bar{x}(T)-x(T)+\dot{x}(T)(\bar{T}-T)) + H_p'(\bar{p}(T)-p(T)+\dot{p}(T)(\bar{T}-T))$$
$$+ H_u'(\bar{u}(T)-u(T)) + \bar{M}_{T\nu}'(\bar{\nu}-\nu) + (H_T(T)+\bar{M}_{TT})(\bar{T}-T)$$
$$= -(H(T)+\bar{M}_T) \quad \ldots(140)$$

$$\psi_x(\bar{x}(T)-x(T)+\dot{x}(T)(\bar{T}-T)) + \psi_T(\bar{T}-T) = -\psi \quad \ldots(141)$$

where $H(T)$, ψ and all the partial derivatives are calculated at the current point (x,p,u,ν,T).

In the generalized Newton-Raphson method ([36]-[38]) (also called the quasilinearization method [39] by Bellman) u is determined from the relation $H_u = 0$ and a linear two-point boundary value problem is posed at each iteration. Since $H_u = 0$, from (138) we get

$$(\bar{u}-u) = -H_{uu}^{-1}(H_{ux}(\bar{x}-x) + H_{up}(\bar{p}-p)) \quad \ldots(142)$$

---

*Modified Newton-Raphson approximation may be employed to all or some of the equations.

Here we have assumed the nonsingularity of H_{uu}. Substituting the value of $(\bar{u}-u)$ from (142) into (136) and (137) we have

$$\dot{\bar{x}} = A\bar{x} + B\bar{p} + v \qquad \ldots\ldots(143)$$

$$\dot{\bar{p}} = C\bar{x} - A'\bar{p} + w \qquad \ldots\ldots(144)$$

where A, B and C are defined as before (see (107)-(109)), $v = -Ax-Bp+H_p$ and $w = -Cx+A'p-H_x$. The equations (139), (140), (141), (143) and (144) together with $x(t_0) = x_0$ constitute the linear two-point boundary value problem which must be solved to get the next estimate of x, p, ν and T.

Let us write (139) - (141) in a simplified form. Substituting the value of $\bar{p}(T)-p(T)+\dot{p}(T)(\bar{T}-T)$ from (139) into (140) and rearranging terms we have

$$(H_x'(T)+\bar{M}_{Tx}'+H_p'\bar{M}_{xx})\bar{x}(T) + (H_p'\bar{M}_{x\nu}+\bar{M}_{T\nu}')\bar{\nu}$$

$$+ \Big((H_x'(T)+\bar{M}_{Tx}'+H_p'\bar{M}_{xx})\dot{x}(T) + (H_p'\bar{M}_{xT}+H_T(T)+\bar{M}_{TT})\Big)\bar{T}$$

$$- (H_x'(T)+\bar{M}_{Tx}'+H_p'\bar{M}_{xx})x(T) - (H_p'\bar{M}_{x\nu}+\bar{M}_{T\nu}')\nu$$

$$- \Big((H_x'(T)+\bar{M}_{Tx}'+H_p'\bar{M}_{xx})\dot{x}(T) + H_p'\bar{M}_{xT}+H_T(T)+\bar{M}_{TT}\Big)T$$

$$- H_p'(p(T)-\bar{M}_x) + H(T) + \bar{M}_T = 0 \qquad \ldots\ldots(145)$$

Let

$$\bar{M}_{xx} = \alpha, \quad \boldsymbol{\gamma}_x' = \beta, \quad (-\dot{p}(T)+\bar{M}_{xx}\dot{x}(T)+\bar{M}_{xT}) = \gamma$$

$$- (\alpha\, x(T)+\beta\boldsymbol{\nu}+\gamma T - \bar{M}_x) = \delta_1, \quad (\boldsymbol{\gamma}_x \dot{x}(T)+\boldsymbol{\gamma}_T) = \eta$$

$$- (\beta'x(T)+\eta T) + \boldsymbol{\mathscr{V}} = \delta_2, \quad (H_x(T)+\bar{M}_{Tx}+\bar{M}_{xx}H_p) = \hat{\gamma}$$

$$(\bar{M}_{x\nu}'H_p+\bar{M}_{T\nu}) = \hat{\eta}, \quad \hat{\gamma}'\dot{x}(T)+H_p'\bar{M}_{xT}+H_T(T)+\bar{M}_{TT} = \mu$$

$$-(\hat{\gamma}'x(T)+\tilde{\eta}\bar{\nu}+\mu T) - H'_p(p(T)-\bar{M}_x) + H(T) + \bar{M}_T = \delta_3$$

Then, equations (139), (141) and (145) can be written as

$$\bar{p}(T) = \alpha \bar{x}(T) + \beta\bar{\nu} + \gamma\bar{T} + \delta_1 \quad \ldots\ldots(146)$$

$$0 = \beta'\bar{x}(T) + \eta\bar{T} + \delta_2 \quad \ldots\ldots(147)$$

$$0 = \hat{\gamma}'\bar{x}(T) + \hat{\eta}'\bar{\nu} + \mu\bar{T} + \delta_3 \quad \ldots\ldots(148)$$

The linear two-point boundary value problem given by $x(t_0) = x_0$ and the equations (143), (144) and (146) - (148) can be solved in several ways. We discuss here the Riccati transformation method. The equations (146) - (148) suggest the following Riccati transformation:

$$\bar{p} = P\bar{x} + R\bar{\nu} + h_1\bar{T} + h \quad \ldots\ldots(149)$$

$$0 = R'\bar{x} + Q\bar{\nu} + g_1\bar{T} + g \quad \ldots\ldots(150)$$

$$0 = \hat{h}'_1\bar{x} + \hat{g}'_1\bar{\nu} + \lambda_1\bar{T} + \lambda \quad \ldots\ldots(151)$$

It is easy to check that the functions $P, R, \ldots, \lambda$ satisfy the following differential equations and the boundary conditions.

$$\dot{P} = -PA - A'P - PBP + C, \qquad P(T) = \alpha \quad \ldots\ldots(152)$$

$$\dot{R} = -(A' + PB)R, \qquad R(T) = \beta \quad \ldots\ldots(153)$$

$$\dot{h}_1 = -(A' + PB)h_1, \qquad h_1(T) = \gamma \quad \ldots\ldots(154)$$

$$\dot{h} = -(A' + PB)h + w - Pv, \qquad h(T) = \delta_1 \quad \ldots\ldots(155)$$

$$\dot{Q} = -R'BR, \qquad Q(T) = 0 \quad \ldots\ldots(156)$$

$$\dot{g}_1 = -R'Bh_1, \qquad g_1(T) = \eta \quad \ldots\ldots(157)$$

$$\dot{g} = R'(Bh + v), \qquad g(T) = \delta_2 \quad \ldots\ldots(158)$$

$$\dot{\hat{h}}_1 = -(A' + PB)\hat{h}_1 \qquad \hat{h}_1(T) = \hat{\gamma} \quad \ldots\ldots(159)$$

$$\dot{\hat{g}}_1 = -R'B\hat{h}_1, \qquad \hat{g}_1(T) = \hat{\eta} \quad \ldots\ldots(160)$$

$$\dot{\lambda}_1 = -\hat{h}'_1 B \hat{h}_1, \qquad \lambda_1(T) = \mu \quad \ldots\ldots(161)$$

$$\dot{\lambda} = \hat{h}_1'(Bh + v), \qquad \lambda(T) = \delta_3 \qquad \ldots\ldots(162)$$

The vector $(\bar{\boldsymbol{v}},\bar{T})$ is determined by solving (150) - (151) at $t = t_0$:

$$\begin{pmatrix} Q(t_0) & g_1(t_0) \\ \hat{g}_1'(t_0) & \lambda_1(t_0) \end{pmatrix} \begin{pmatrix} \bar{\boldsymbol{v}} \\ \bar{T} \end{pmatrix} = - \begin{pmatrix} R'(t_0)x_0 + g(t_0) \\ \hat{h}_1'(t_0)x_0 + \lambda(t_0) \end{pmatrix} \ldots(163)$$

We note that the application of the modified Newton-Raphson approximation to the equations $\Omega = 0$ and $\Psi = 0$ give the equations

$$\begin{pmatrix} Q(t_0) & g_1(t_0) \\ \hat{g}_1'(t_0) & \lambda_1(t_0) \end{pmatrix} \begin{pmatrix} \bar{\boldsymbol{v}} \\ \bar{T} \end{pmatrix} = \begin{pmatrix} \varepsilon_1 \Psi - R'(t_0)x_0 - g(t_0) \\ \varepsilon_2 \Omega - \hat{h}_1'(t_0)x_0 - \lambda(t_0) \end{pmatrix} \ldots(164)$$

for the determination of $(\bar{\boldsymbol{v}},\bar{T})$ where $0 \leq \varepsilon_1, \varepsilon_2 < 1$.

Having obtained $(\bar{\boldsymbol{v}},\bar{T})$, the new estimate of x is determined first, by integrating

$$\dot{\bar{x}} = (A+BP)\bar{x} + B(R\bar{\boldsymbol{v}}+h_1\bar{T}+h)+v, \qquad \bar{x}(t_0) = x_0 \qquad \ldots\ldots(165)^*$$

forward from t_0 to $\bar{T}$ and then the new estimate of p is computed from (149), where equation (165) is obtained by eliminating $\bar{p}$ from (143) by means of (149). The process ends when the equations $\dot{x} - H_p = 0$ and $\dot{p} + H_x = 0$ (with u determined from $H_u = 0$) and the equations $\zeta = 0$, $\Omega = 0$ and $\Psi = 0$ are satisfied.

The computer algorithm might proceed as follows:

---

*If $\bar{T} > T$, then the variable terms $(A+BP)$ and $B(R\bar{\boldsymbol{v}}+h_1\bar{T}+h)+v$ appearing in (165) must be extrapolated over the interval $(T,\bar{T}]$.

(i) Pick a point (x, p, ν, T) with $x(t_0) = x_0$.

(ii) Solve the differential equations (152) - (155) and (159) backward in time from T to t_0 and store the solutions so obtained.

(iii) Integrate (156) - (158) and (160) - (162) and determine the new estimates of ν and T from (163) and (164).

(iv) Determine the new estimates of x and p from (165) and (149) respectively.

(v) Repeat steps (ii) - (iv) if

$$\sum_{i=1}^{n} \{ \max_{0 \leq t \leq \tilde{T}} |\bar{x}_i(t) - x_i(t)| + \max_{0 \leq t \leq \tilde{T}} |\bar{p}_i(t) - p_i(t)| \}$$

is not less than a preassigned small positive number ε, where $\tilde{T} = \min(\bar{T}, T)$; otherwise proceed to step (vi).

(vi) As a final check solve the original nonlinear equations $\dot{x} = H_p$ and $\dot{p} = -H_x$ (where u is determined from $H_u = 0$) with $x(t_0) = x_0$ and $p(t_0) = \bar{p}(t_0)$ determined from (149). If the equations $\zeta = 0$, $\Omega = 0$ and $\Psi = 0$ are satisfied to some desired tolerances the process is terminated; otherwise, steps (ii) to (v) are repeated by picking another ε in step (v).

We mention that in the fixed terminal time problem the Riccati transformation is given by (149) and (150) with $h_1 = 0$ and $g_1 = 0$ and that the variable terminal time problem may be solved as a sequence of fixed terminal time problems (see Section 4.1).

Finally, we consider the control problem with constraints on the control variables. This problem can be solved either by applying the penalty function technique or by introducing variables ξ_j such that

$$R_j \equiv (u_j - \alpha_j)(\beta_j - u_j) = \xi_j^2, \quad j=1,\ldots,m \quad \ldots(166)$$

The introduction of the equality constraints (166) results in the replacement of the condition $H_u = 0$ by $H_u + R_u' \mu = 0$ where μ, an m-dimensional vector function, satisfies the equations $\mu_j \xi_j = 0$, $j = 1,\ldots,m$ (see e.g. [36]). The problem is now solved by applying the Newton-Raphson approximation to the equations $\dot{x} - H_p = 0$, $\dot{p} + H_x = 0$, $H_u + R_u' \mu = 0$, $\zeta = 0$, $\gamma = 0$, $\Omega = 0$, $R - \xi^2 = 0$ and $\mu_j \xi_j = 0$, $j = 1,\ldots,m$. We omit the detail here and refer to Kenneth and McGill*[36] for a further discussion of this approach.

8.0 CONCLUDING REMARKS

The methods discussed here (or their variants) have been compared by some authors (see e.g. [4], [22] and [40]) and it is generally agreed that no one method is better (in the sense of the simplicity of formulation, convergence, computer time and computer storage) than the others in all situations. Each method, therefore, must be judged in the light of the problem at hand. For the purpose of constructing a general optimization technique it may be advantageous to use a combination of two methods in such a way that we get a rapid initial convergence and a rapid final convergence. This may be achieved, for example, by starting the computation with the conjugate gradient method and ending with the successive sweep method or the generalized Newton-Raphson method.

---

*Note that Kenneth and McGill used the method of complementary functions for solving linear two-point boundary value problems.

In discussing the computational methods, the emphasis has been on the basic points of each method and not all the possible variants of each method have been considered. It is, however, hoped that by studying the methods discussed here the 'optimal controllist' will be able to modify them and also will be able to discover (or rediscover) by himself some new methods.

REFERENCES

1. R. E. Kalman, P. L. Falb and M. A. Arbib, Topics in Mathematical System Theory, Chapter 5, McGraw-Hill, 1969.

2. M. Athans, The Status of Optimal Control Theory and Applications for Determnistic Systems, IEEE Trans. Automatic Control, Vol. AC-11, No. 3, July 1966, pp. 580-596.

3. A. V. Balakrishnan and L. W. Neustadt (Eds.), Computing Methods in Optimization Problems, Academic Press, 1964.

4. R. E. Kopp and H. G. Moyer, Trajectory Optimization Techniques, Advances in Control Systems, Vol. 4, (Ed. C.T. Leondes) Academic Press 1966, pp. 103-155.

5. R. E. Bellman and S. E. Dreyfus, Applied Dynamic Programming, Princeton University Press, 1962.

6. R. E. Larson, A Survey of Dynamic Programming Computational Procedures, IEEE Trans. Automatic Control, Vol. AC-12, No. 6, December 1967, pp. 767-774.

7. D. H. Jacobson and D. Q. Mayne, Differential Dynamic Programming, American Elsevier, 1969.

8. L. D. Berkovitz, Variational Methods in Problems of Control and Programming, Journal Math. Anal. Appl., 3, 1961, pp. 145-169.

9. M. R. Hestenes, Calculus of Variations and Optimal Control Theory, John Wiley and Sons, Inc. 1966.

10. J. W. Sutherland and E. V. Bohn, A Numerical Trajectory Optimization Method Suitable for a Computer of Limited Memory, IEEE Trans. Automatic Control, Vol. AC-11, No. 3, July 1966, pp. 440-447.

11. H. J. Kelley, Methods of Gradients in "Optimization Techniques" (G. Leitmann Ed.) Academic Press 1962; pp. 205-254.

12. L. S. Lasdon, S. K. Mitter and A. D. Waren, The Conjugate Gradient Method for Optimal Control Problems, IEEE Trans. Auto. Cont., Vol. AC-12, No. 2, April 1967, pp. 132-138.

13. J. F. Sinnott, Jr. and D. G. Luenberger, Solution of Optimal Control Problems by the Method of Conjugate Gradients, Proc. Joint Automatic Control Conf., 1967, University of Pennsylvania, Philadelphia, pp. 566-574.

14. B. Pagurek and C. M. Woodside, The conjugate Gradient Method for Optimal Control Problems with Bounded Control Variables, Automatica, Vol. 4, 1968, pp. 337-349.

15. S. S. Tripathi and K. S. Narendra, Optimization using Conjugate Gradent Methods Dunham Lab. Tech. Rep. CT-27, Dept. EE and Appl. Sc., Yale University.

16. A. E. Bryson and W. F. Denham, A Steepest Ascent Method for Solving Optimum Programming Problems, J. Appl. Mech. (Trans. ASME, Ser. E) 29, June 1962, pp. 247-257.

17. L. S. Lasdon, A. D. Waren and R. K. Rice, An Interior Penalty Method for Inequality Constrained Optimal Control Problems, IEEE Trans. Auto. Cont. Vol. AC-12, No. 4, Aug. 1967, pp. 388-395.

18. S. R. McReynolds and A. E. Bryson, A Successive Sweep Method for Solving Optimal Programming Problems, Preprints, 1965 JACC (Troy, N.Y.) pp. 551-555.

19. W. E. Schmitendorf and S. J. Citron, On the Applicability of the Sweep Method to Optimal Control Problems, IEEE Trans. Automatic Control, Vol AC-14, No. 1, Feb., 1969, pp. 69-72.

20. H. J. Kelley, R. E. Kopp and H. G. Moyer, A Trajectory Optimization Technique Based upon the Theory of the Second Variation, in "Progress in Astronautics and Aeronautics: Celestial Mechanics and Astrodynamics" (V.G. Szebehely Ed.), Vol. 14, p. 559, Academic Press, New York, 1964.

21. C. W. Merriam, An Algorithm for the Iterative Solution of a Class of Two-point Boundary Value Problems. SIAM Journal Control, Ser. A, Vol. 2, No. 1, 1964, pp. 1-10.

22. S. K. Mitter, Successive Approximation Methods for the Solution of Optimal Control Problems, Automatica, Vol. 3, 1966, pp. 135-149.

23. I. H. Mufti, Initial-value Methods for Two-point Boundary-value Problems, National Research Council of Canada, Div. of Mech. Eng'g., Report 1969 (in print).

24. U. Hochstrasser, Numerical Methods for Finding Solutions of Nonlinear Equations, Chapter 7, Survey of Numerical Analysis, Edited by J. Todd, McGraw-Hill, 1962.

25. A. S. Householder, Principles of Numerical Analysis, McGraw-Hill, 1953.

26. F. J. Zeleznik, Quasi-Newton Methods for Nonlinear Equations, Journal ACM, Vol. 15, No. 2, April 1968, pp. 265-271.

27. J. V. Breakwell, J. L. Speyer and A. E. Bryson, Optimization and Control of Nonlinear Systems using the Second Variation, J. SIAM Control Ser. A, Vol. 1, No. 2, 1963, pp. 193-223.

28. A. H. Jazwinski, Optimal Trajectories and Linear Control of Nonlinear Systems, AIAA Journal, Vol. 2, No. 8, Aug. 1964, pp. 1371-1379.

29. M. D. Levine, Trajectory Optimization Using the Newton-Raphson Method, Automatica, Vol. 3, 1966, pp. 203-217.

30. C. H. Knapp and P. A. Frost, Determination of Optimum Control and Trajectories Using the Maximum Principle in Association with a Gradient Technique, IEEE Trans. Automatic Control, Vol AC-10, No. 2, April 1965, pp. 189-193.

31. L. G. Birta and P. J. Trushel, The TEF/Davidon-Fletcher-Powell Method in the Computation of Optimal Controls, Proceedings JACC 1969, Boulder, Colorado, pp. 259-266.

32. H. A. Spang III, A Review of Minimization Techniques for Nonlinear Functions, SIAM Review, Vol. 4, Oct. 1962, pp. 343-365.

33. M. J. D. Powell and R. Fletcher, A Rapidly Convergent Descent Method for Minimization, The Computer Journal, Vol. 6, 1963, pp. 163-168.

34. B. V. Shah, R. J. Buehler and O. Kempthorne, Some Algorithms for Minimizing a Function of Several Variables, J. SIAM, Vol. 12, No. 1, March 1964, pp. 74-92.

35. C. G. Broyden, Quasi-Newton Methods and Their Application to Function Minimization, Math. Comput. Vol. 21, July 1967, pp. 368-381.

36. P. Kenneth and R. McGill, Two-point Boundary Value Problem Techniques, in C.T. Leondes (Ed.) Advances in Control Systems, Vol. 3, Academic Press, New York, 1966, pp. 69-109.

37. C. H. Schley and I. Lee, Optimal Control Computation by the Newton-Raphson Method and the Riccati Transformation, IEEE Trans. Aut. Cont. Vol. AC-12, April, 1967, pp. 139-144.

38. A. G. Longmuir, Numerical Computation of Nearly-optimal Feedback Control Laws and Optimal Control Programs, Ph.D. Thesis, Dept. of E.E., University of British Columbia, Vancouver, B.C. 1968.

39. R. E. Bellman and R. E. Kalaba, Quasilinearization and Nonlinear Boundary-value Problems, Elsevier, Amsterdam, 1965.

40. J. M. Lewallen and B. D. Tapley, Analysis and Comparison of Several Numerical Optimization Methods, AIAA 5th Aerospace Sciences Meeting, paper No. 67-58, Jan. 23-26, 1967.

ADDITIONAL REFERENCES

D. Isaacs, C. T. Leondes and R. A. Niemann, On A Sequential Optimization Approach in Nonlinear Control, Proc. JACC, Seattle, Washington, 1966, pp. 158-166.

T. E. Bullock and G. F. Franklin, A Second-order Feedback Method for Optimal Control Computations, IEEE Trans. Auto. Control, Vol. AC-12, No. 6, Dec. 1967, pp. 666-673.

G. J. Lastman, A Modified Newton's Method for Solving Trajectory Optimization Problems, AIAA Journal, Vol. 6, No. 5, May 1968, pp. 777-780.

C. T. Leondes and G. Paine, Extensions in Quasilinearization Techniques for Optimal Control, Journal of Optimization Theory and Applications, Vol. 2, No. 5, 1968, pp. 316-330.

A.V. Balakrishnan, On a New Computing Technique in Optimal Control, SIAM J. on Control, Vol. 6, No. 2, May 1968, pp. 149-173.

L. W. Taylor, H. J. Smith and K. W. Iliff, Experience using Balakrishnan's Epsilon Technique to Compute Optimum Flight Profiles, AIAA 7th Aerospace Sciences Meeting, New York, Jan. 20-22, 1969, AIAA Paper No. 69-75.

P. Dyer and S. R. McReynolds, Optimization of Control Systems with Discontinuities and Terminal Constraints, IEEE Trans. Auto. Control, Vol. AC-14, No. 3, June 1969, pp. 223-229.

L. B. Horwitz and P. E. Sarachik, A Survey of Two Recent Iterative Techniques for Computing Optimal Control Signals, Proceedings 4th Congress IFAC, Warsaw, Poland, June 1969.

Lecture Notes in Operations Research and Mathematical Systems

Vol. 1: H. Bühlmann, H. Loeffel, E. Nievergelt, Einführung in die Theorie und Praxis der Entscheidung bei Unsicherheit. 2. Auflage, IV, 125 Seiten 4°. 1969. DM 12,– / US $ 3.30

Vol. 2: U. N. Bhat, A Study of the Queueing Systems M/G/1 and GI/M/1. VIII, 78 pages. 4°. 1968. DM 8,80 / US $ 2.50

Vol. 3: A. Strauss, An Introduction to Optimal Control Theory. VI, 153 pages. 4°. 1968. DM 14,– / US $ 3.90

Vol. 4: Einführung in die Methode Branch and Bound. Herausgegeben von F. Weinberg. VIII, 159 Seiten. 4°. 1968. DM 14,– / US $ 3.90

Vol. 5: L. Hyvärinen, Information Theory for Systems Engineers. VIII, 295 pages. 4°. 1968. DM 15,20 / US $ 4.20

Vol. 6: H. P. Künzi, O. Müller, E. Nievergelt, Einführungskursus in die dynamische Programmierung. IV, 103 Seiten. 4°. 1968. DM 9,– / US $ 2.50

Vol. 7: W. Popp, Einführung in die Theorie der Lagerhaltung. VI, 173 Seiten. 4°. 1968. DM 14,80 / US $ 4.10

Vol. 8: J. Teghem, J. Loris-Teghem, J. P. Lambotte, Modèles d'Attente M/G/1 et GI/M/1 à Arrivées et Services en Groupes. IV, 53 pages. 4°. 1969. DM 6,– / US $ 1.70

Vol. 9: E. Schultze, Einführung in die mathematischen Grundlagen der Informationstheorie. VI, 116 Seiten. 4°. 1969. DM 10,– / US $ 2.80

Vol. 10: D. Hochstädter, Stochastische Lagerhaltungsmodelle. VI, 269 Seiten. 4°. 1969. DM 18,– / US $ 5.00

Vol. 11/12: Mathematical Systems Theory and Economics. Edited by H. W. Kuhn and G. P. Szegö. VIII, IV, 486 pages. 4°. 1969. DM 34,– / US $ 9.40

Vol. 13: Heuristische Planungsmethoden. Herausgegeben von F. Weinberg und C. A. Zehnder. II, 93 Seiten. 4°. 1969. DM 8,– / US $ 2.20

Vol. 14: Computing Methods in Optimization Problems. Edited by A. V. Balakrishnan. V, 191 pages. 4°. 1969. DM 14,– / US $ 3.90

Vol. 15: Economic Models, Estimation and Risk Programming: Essays in Honor of Gerhard Tintner. Edited by K. A. Fox, G. V. L. Narasimham and J. K. Sengupta. VIII, 461 pages. 4°. 1969. DM 24,– / US $ 6.60

Vol. 16: H. P. Künzi und W. Oettli, Nichtlineare Optimierung: Neuere Verfahren, Bibliographie. IV, 180 Seiten. 4°. 1969. DM 12,– / US $ 3.30

Vol. 17: H. Bauer und K. Neumann, Berechnung optimaler Steuerungen, Maximumprinzip und dynamische Optimierung. VIII, 188 Seiten. 4°. 1969. DM 14,– / US $ 3.90

Vol. 18: M. Wolff, Optimale Instandhaltungspolitiken in einfachen Systemen. V, 143 Seiten. 4°. 1970. DM 12,– / US $ 3.30

Vol. 19: L. Hyvärinen, Mathematical Modeling for Industrial Processes. VI, 122 pages. 4°. 1970. DM 10,– / US $ 2.80

Vol. 20: G. Uebe, Optimale Fahrpläne. IX, 161 Seiten. 4°. 1970. DM 12,– / US $ 3.30

Vol. 21: Th. Liebling, Graphentheorie in Planungs- und Tourenproblemen am Beispiel des städtischen Straßendienstes. IX, 118 Seiten. 4°. 1970. DM 12,– / US $ 3.30

Vol. 22: W. Eichhorn, Theorie der homogenen Produktionsfunktion. VIII, 119 Seiten. 4°. 1970. DM 12,– / US $ 3.30

Vol. 23: A. Ghosal, Some Aspects of Queueing and Storage Systems. IV, 93 pages. 4°. 1970. DM 10,– / US $ 2.80

Vol. 24: Feichtinger, Lernprozesse in stochastischen Automaten.
V, 66 Seiten. 4°. 1970. DM 6,– / $ 1.70

Vol. 25: R. Henn und O. Opitz, Konsum- und Produktionstheorie I.
II, 124 Seiten. 4°. 1970. DM 10,– / $ 2.80

Vol. 26: D. Hochstädter und G. Uebe, Ökonometrische Methoden.
XII, 250 Seiten. 4°. 1970. DM 18,– / $ 5.00

Vol. 27: I. H. Mufti, Computational Methods in Optimal Control Problems.
IV, 45 pages. 4°. 1970. DM 6,– / $ 1.70

Beschaffenheit der Manuskripte

Die Manuskripte werden photomechanisch vervielfältigt; sie müssen daher in sauberer Schreibmaschinenschrift geschrieben sein. Handschriftliche Formeln bitte nur mit schwarzer Tusche eintragen. Notwendige Korrekturen sind bei dem bereits geschriebenen Text entweder durch Überkleben des alten Textes vorzunehmen oder aber müssen die zu korrigierenden Stellen mit weißem Korrekturlack abgedeckt werden. Falls das Manuskript oder Teile desselben neu geschrieben werden müssen, ist der Verlag bereit, dem Autor bei Erscheinen seines Bandes einen angemessenen Betrag zu zahlen. Die Autoren erhalten 75 Freiexemplare.

Zur Erreichung eines möglichst optimalen Reproduktionsergebnisses ist es erwünscht, daß bei der vorgesehenen Verkleinerung der Manuskripte der Text auf einer Seite in der Breite möglichst 18 cm und in der Höhe 26,5 cm nicht überschreitet. Entsprechende Satzspiegelvordrucke werden vom Verlag gern auf Anforderung zur Verfügung gestellt.

Manuskripte, in englischer, deutscher oder französischer Sprache abgefaßt, nimmt Prof. Dr. M. Beckmann, Department of Economics, Brown University, Providence, Rhode Island 02912/USA oder Prof. Dr. H.P. Künzi, Institut für Operations Research und elektronische Datenverarbeitung der Universität Zürich, Sumatrastraße 30, 8006 Zürich entgegen.

Cette série a pour but de donner des informations rapides, de niveau élevé, sur des développements récents en économétrie mathématique et en recherche opérationnelle, aussi bien dans la recherche que dans l'enseignement supérieur. On prévoit de publier

1. des versions préliminaires de travaux originaux et de monographies
2. des cours spéciaux portant sur un domaine nouveau ou sur des aspects nouveaux de domaines classiques
3. des rapports de séminaires
4. des conférences faites à des congrès ou à des colloquiums

En outre il est prévu de publier dans cette série, si la demande le justifie, des rapports de séminaires et des cours multicopiés ailleurs mais déjà épuisés.

Dans l'intérêt d'une diffusion rapide, les contributions auront souvent un caractère provisoire; le cas échéant, les démonstrations ne seront données que dans les grandes lignes. Les travaux présentés pourront également paraître ailleurs. Une réserve suffisante d'exemplaires sera toujours disponible. En permettant aux personnes intéressées d'être informées plus rapidement, les éditeurs Springer espèrent, par cette série de »prépublications«, rendre d'appréciables services aux instituts de mathématiques. Les annonces dans les revues spécialisées, les inscriptions aux catalogues et les copyrights rendront plus facile aux bibliothèques la tâche de réunir une documentation complète.

Présentation des manuscrits

Les manuscrits, étant reproduits par procédé photomécanique, doivent être soigneusement dactylographiés. Il est recommandé d'écrire à l'encre de Chine noire les formules non dactylographiées. Les corrections nécessaires doivent être effectuées soit par collage du nouveau texte sur l'ancien soit en recouvrant les endroits à corriger par du verni correcteur blanc.

S'il s'avère nécessaire d'écrire de nouveau le manuscrit, soit complètement, soit en partie, la maison d'édition se déclare prête à verser à l'auteur, lors de la parution du volume, le montant des frais correspondants. Les auteurs recoivent 75 exemplaires gratuits.

Pour obtenir une reproduction optimale il est désirable que le texte dactylographié sur une page ne dépasse pas 26,5 cm en hauteur et 18 cm en largeur. Sur demande la maison d'édition met à la disposition des auteurs du papier spécialement préparé.

Les manuscrits en anglais, allemand ou francais peuvent être adressés au Prof. Dr. M. Beckmann, Department of Economics, Brown University, Providence, Rhode Island 02912/USA ou au Prof. Dr. H.P. Künzi, Institut für Operations Research und elektronische Datenverarbeitung der Universität Zürich, Sumatrastraße 30, 8006 Zürich.

MIX
Papier aus verantwortungsvollen Quellen
Paper from responsible sources
FSC® C105338

If you have any concerns about our products,
you can contact us on
ProductSafety@springernature.com

In case Publisher is established outside the EU,
the EU authorized representative is:
**Springer Nature Customer Service Center GmbH
Europaplatz 3, 69115 Heidelberg, Germany**

Printed by Libri Plureos GmbH
in Hamburg, Germany